D0399420

the Philadelphia Chromosome

the Philadelphia Chromosome

A Mutant Gene and the Quest to Cure Cancer at the Genetic Level

JESSICA WAPNER

FOREWORD BY ROBERT A. WEINBERG, PHD

The Philadelphia Chromosome:
A Mutant Gene and the Quest to Cure Cancer at the Genetic Level

The Experiment, LLC
260 Fifth Avenue
New York, NY 10001–6408
www.theexperimentpublishing.com

The Experiment's books are available at special discounts when purchased in bulk for premiums and sales promotions as well as for fund-raising or educational use. For details, contact us at info@theexperimentpublishing.com.

Library of Congress CIP data
Wapner, Jessica.
The Philadelphia chromosome : a mutant gene and the quest to cure cancer at the genetic level / Jessica Wapner ; foreword by Robert A. Weinberg.
p. ; cm.
Includes bibliographical references and index.
ISBN 978-1-61519-067-6 (pbk.) -- ISBN 978-1-61519-165-9 (ebook)
I. Title.
[DNLM: 1. Leukemia, Myelogenous, Chronic, BCR-ABL Positive--history--United States. 2. Philadelphia Chromosome--United States. 3. Fusion Proteins, bcr-abl--history--United States. 4. Genetic Therapy--history--United States. QZ 11 AA1]

616.99'419042--dc23
2012047686

ISBN 978-1-61519-067-6
Ebook ISBN 978-1-61519-165-9

Cover design by Jason Gabbert
Cover photograph © 2006 Peter C. Nowell, MD, Department of Pathology and Clinical Laboratory of the University of Pennsylvania School of Medicine
Author photograph © Meredith Heuer
Text design by Pauline Neuwirth, Neuwirth & Associates, Inc.

Manufactured in the United States of America
Distributed by Workman Publishing Company, Inc.
Distributed simultaneously in Canada by Thomas Allen & Son Ltd.
First published May 2013
10 9 8 7 6 5 4 3 2

To Evangelos and my other guiding lights

| CONTENTS |

AUTHOR'S NOTE

This book draws extensively from interviews I conducted with those who lived this story. Unless otherwise indicated, all quotes herein are from those interviews, which took place in 2012 or, in the case of a few scientists, 2007. Where quotes are taken from other material, the original source is noted.

Also, chronic myeloid leukemia (CML), the blood cancer that is this book's focus, is alternately referred to as chronic myelogenous leukemia. I elected to refer to chronic myeloid leukemia, except when I quote from interviews or previously published material in which "myelogenous" was used; in these instances, "myelogenous" has been retained.—*JW*

FOREWORD

By Robert A. Weinberg, PhD

A widespread illusion is that cancer is a disease of modernity, an artifact of pollution and bad diet and myriad other factors associated with a modern lifestyle. The truth is quite different: The disease of cancer threatens all multicellular life with greater or lesser frequency. In the case of our own species, cancer incidence has exploded because we now live long enough to develop a disease—much like Alzheimer's—that largely strikes the aged.

Until recently, we did not how and why the disease arose, and yet, in spite of this, we developed the means to treat some—but hardly all—forms of the disease. The use of chemotherapy and radiation has had a remarkable effect in treating some cancers and almost no effect on others. By the early 1970s, however, it was already clear that these cytotoxic therapies had yielded almost as much benefit as they possibly could; that is, they had exhausted their potential for making major inroads into reducing cancer-associated mortality. Those looking over the scientific horizon concluded that new ways of treating the disease were required.

The thinking of those interested in such innovations in cancer treatment focused on how the disease was being caused. If only one could understand the defects within cancer cells, they reasoned, novel ways of treating the disease would surely emerge. This article of faith gained currency with the discoveries in the second half of the 1970s that distinct cancer-causing genes could be found within cancer cells. The genes—soon called oncogenes—appeared to be the motive forces behind the runaway proliferation of many types of human cancer cells. By attacking these genes (and more specifically, the proteins that they produced), highly targeted, extremely effective therapies could be developed, or so the thinking went.

The proponents of this new approach to dealing with cancer portrayed their strategy as a means of developing anticancer drugs rationally, by focusing on specific molecular defects within cancer cells. They contrasted this new way of treating cancer with the traditional strategies involving cytotoxic treatments, which were portrayed as blunt, crude instruments that had been used with limited success for the previous several decades. The latter therapies had been developed without any knowledge whatsoever of why cancer cells behaved aberrantly. For the first time, the prospect was bright; by attacking malfunctioning proteins within cancer cells, tumors as a whole could be brought to their knees.

By the early 1980s, the research into the molecular origins of cancer began an explosive growth, and by the end of the century we had amassed a truly extraordinary body of detailed information on why cancer cells proliferate abnormally. Translating these results into novel treatments, however, has not come so easily. There have been the usual scientific obstacles of developing drugs that could strike the cancer cells with great specificity, selectively killing those cells while leaving their normal counterparts relatively unaffected. Such selectivity in targeting cancer cells is rarely absolute; almost always, side-effect toxicities attend even the most successful targeted therapies. Then there were the economic considerations, specifically, whether the rapidly growing cost of developing novel anticancer therapeutics could ever be recouped through clinical drug treatments.

The Philadelphia Chromosome focuses on what is widely viewed (at least among those of us in cancer research and treatment) as the "poster child" of rational drug development. The story that this book relates vindicates those who dreamed that one day cancers could be treated through rationally designed drugs—and in fact remains the major success story to this day. It has been followed, disappointingly, by few comparable success stories. It is also a story that almost didn't happen.

The development of Gleevec might never have occurred because those keeping the books in the drug company that incubated this drug argued that the disease was too uncommon to justify major investments in its development. They said that the costs of drug development would never be compensated by sales. Fortunately they did not prevail, and the prophets of rational drug design pushed ahead and generated a truly spectacular therapeutic agent.

By following the step-by-step process launched by the discovery of an aberrant chromosome in the leukemic cells of those suffering from chronic myeloid leukemia, it becomes possible to see how truly challenging anticancer drug development can be, both at the preclinical level (i.e., in the laboratory) and in the clinic itself. The diagnosis of CML was once a death sentence because of its almost-inevitable progression after several years to the "blast crisis" that signals the end of life for patients. These days, CML diagnosed early can become a chronic disease—never cured, strickly speaking, but kept in check so effectively that patients often forget they harbor small nests of leukemic cells in their marrow—cells that are kept penned up by the miracle drug that they regularly take.

Why has Gleevec been so spectacularly successful in treating CML, while analogous drugs treating other forms of cancer have had mixed results, often eliciting short-term responses followed by drug-resistant clinical relapses? The answers lie in part in the fortuitous choice of treating a disease that has not yet become aggressive, stopping it in its tracks, in contrast to most other forms of cancer, which are frequently diagnosed late in their natural progression and thus endowed with the means of evading the therapies dispatched to eliminate them. This book makes good reading for those interested in the work of the heroes who pushed this drug forward to its truly brilliant successes.

ROBERT A. WEINBERG, PHD, is a Daniel K. Ludwig Professor for Cancer Research at MIT, an American Cancer Society Research Professor, and a member of the US National Academy of Sciences. He is an internationally recognized authority on the genetic basis of human cancer. Credited with the discovery of the first human cancer-causing gene as well as the first tumor suppressor gene, he was awarded the National Medal of Science in 1997 and the Wolf Prize in Medicine in 2004. He is the author or editor of five books and more than 325 articles.

the Philadelphia Chromosome

PRELUDE

Down to the Bone

February 2012

Gary Eichner sat in a chair backed up against a wall. Across the room, his nurse was half hidden by a computer. She scrolled through his medical chart on a monitor he couldn't see, typing his responses to her questions. He answered and smiled as if he were just fine, as though that would somehow make it so. He kept his worry concealed, silently wondering what the next hour would reveal about the disease that had so suddenly overtaken his life. Fear permeated his every thought.

The nurse ran through the litany of side effects that he might be experiencing from the leukemia medication he'd recently begun to take. "Any chest pain?" she asked. "Heart problems? Any swelling in your ankles? Any problems with nausea?"

Eichner's blanket "no" was tempered only by his description of what happened when he took the drug on an empty stomach. "It's massive cramping, massive pain in the stomach," he told the nurse. "It's just like the worst thing you've ever had."

The nurse knew that Eichner really had no idea of the worst thing someone with his type of leukemia could have. Few patients with Eichner's disease today will ever know that kind of pain.

But if the nurse was thinking anything of the sort, she kept it to herself. After all, her patient was about to have a large, hollow needle inserted into his bone so that his doctor could extract a sample of his marrow. Having witnessed countless such procedures during the twelve years she'd been a nurse to patients taking the drug, she knew how much was riding on this biopsy, and she knew that Eichner, age 43, was well aware that his life was at stake.

If anyone could set him at ease about that, it was his doctor, Brian Druker. "All I care about is that you're making some progress," Druker told his patient, who had come to Druker's clinic in Portland, Oregon, for a bone marrow biopsy. The procedure would reveal whether the medicine—a pill he'd been taking daily for six months—was tackling the leukemia that had invaded his body. Eichner had chronic myeloid leukemia, or CML, a cancer of the white blood cells that, though slow growing, could be fatal. "I will be happy if you're eighteen out of twenty." Eichner nodded his understanding, his jittery foot the only sign of his nervousness.

For Eichner, those numbers were part of the new language he'd learned since his diagnosis in the summer of 2011. As with so many cancer diagnoses, the education began suddenly and unexpectedly. After a day or so of sharp, excruciating kidney pain, Eichner, a single parent of a teenage boy, drove himself to the emergency room at his local hospital in Olympia, Washington, where he'd been living at the time. His sister-in-law, a trauma flight nurse, had told him she thought he had kidney stones, so Eichner was expecting to be admitted for a standard, albeit painful, procedure. But when several doctors walked into his hospital room together, he knew something wasn't right. "You don't have kidney stones," they told him. "You've got what we believe is leukemia."

A blood test soon revealed that Eichner, in otherwise perfect health, had CML. The excess number of white blood cells contained in the sample confirmed that much. The doctors explained that the stabbing pain he'd thought came from his kidneys probably was from his spleen, enlarged by the high concentration of leukemia cells within. Although the slow-moving nature of the disease meant he wasn't in immediate danger, there wasn't any time to lose. For the best chance at long-term survival, he needed to start treatment right away.

As Eichner was rapidly informed, danger could eventually come. If the treatments didn't work, within five years his bone marrow would fill with blast cells, white blood cells that fail to mature and thus are both overly abundant and thoroughly useless. His blood, once free flowing, would turn into viscous sludge. The supply of iron-rich red blood cells that carry oxygen around the body would steadily plummet, leaving him fatigued and anemic, while a decrease in the number of platelets would render his blood unable to clot. As the disease moved from the accelerated phase (more than 15 percent blast cells) to blast crisis (more than 30 percent blast cells), the minuscule capillaries leading to his eyes and brain would clog. His spleen would likely become profoundly enlarged. As his body began shutting down, he would bleed in his brain, in his intestines, and out of every orifice.

Two days later, Eichner was having his first bone marrow biopsy, performed by a nurse who spoke broken English and had to climb on top of Eichner to hammer in the four-inch-long needle to break through his bones, which had become hardened and inflamed from the profusion of white blood cells inside.

Finally, she managed to extract an ounce of marrow, the spongy matter in the center of our bones where new blood cells are made. In the genetics laboratory, a fluorescent dye applied to the coiled strands of DNA inside Eichner's blood cells revealed the telltale sign of CML: a genetic mutation known as the "Philadelphia chromosome," an abnormal chromosome regarded as the defining feature of this fatal, spontaneously arising cancer and discovered more than fifty years ago. Twenty out of twenty cells in a sample of his marrow contained this genetic error.

A family friend with the same type of cancer told Eichner to call a doctor named Brian Druker right away. "Don't do anything before you see him," the friend told Eichner a couple of days after his diagnosis. Three days later, Eichner got a call from Druker, who'd mysteriously—Eichner didn't know exactly how—gotten his phone number. During the next twenty minutes, Druker talked Eichner down from his panic, assuring him that, being in the early stages of the disease, Eichner could wait a week or two before making the trip south to Oregon Health and Science University (OHSU), where Druker had been treating and researching leukemia since 1993. He gave Eichner his professional blessing to go drink beer. It was August, and Eichner had been getting ready to leave for a day at Olympia's annual summer Brew Fest, his brother's idea for getting his mind off leukemia for a while, when the phone rang.

Ten days later, Eichner was on his way to Portland. Within three weeks of his diagnosis, he was swallowing his first tablet of medication, which Druker had been instrumental in developing: a drug aimed at tackling the cancer at its root.

The first few days of treatment were pure hell. Eichner vomited several times a day, often through the night, and the nausea kept him from sleeping. As the drug flushed out the overload of blood cells that had accumulated in his marrow, the pain of his bones readjusting was excruciating. The lack of sleep and collection of medications to help his body cope with this new situation left him weak and pale. Though he continued his supervising work at the new canola oil plant his employer was building, walking the site for hours a day to ensure that the electrical work was correct, he'd be dead tired by early afternoon. Only later, when his cheeks were rosy again, did his crew confess how shockingly sick he'd looked. A visit to Eichner's ex-wife in Colorado meant his teenage son was spared the sight of his work-hard, play-hard father curled up in pain and retching over the toilet. By the time his son returned home, the side effects were over, and Eichner was adjusting to his new reality. Though his muscles were still recovering, he was back at the gym and able to walk the construction site all day without tiring.

Now, IN THE chill of February, the summer before his diagnosis seemed like a dream. He was living in Vancouver now, a work relocation that brought him just twenty minutes away from OHSU. He and his wife, having put their divorce on hold when he was first diagnosed, were now legally separated. Six months after he became a cancer patient, it was time for another bone marrow biopsy to find out if the drug was working. While Eichner's son and brother hung out at the hospital café, Druker explained to Eichner what he was looking for at this stage of treatment. The bone marrow biopsy would provide a trained technician with a sample of marrow cells—twenty of them, to be exact—to examine under a microscope. When Eichner was diagnosed, all twenty cells had the telltale genetic sign of CML. Because the cancer progressed very slowly over several years, even a minor decrease in the number of cells containing the mutant gene would satisfy Druker that the treatment was working. Thus his reassuring comment, "I will be happy if you're eighteen out of twenty." As Druker, his calm blue eyes matching his gentle demeanor, explained to his patient, just two fewer abnormal cells were enough to signal that the medicine was working. He told Eichner how varied the results could be at this stage. There was a chance the drug hadn't done anything, leaving Eichner at twenty out of twenty. Or, he could be well below eighteen, even at zero.

With his black hoodie, jeans, trim goatee, and rugged face, Eichner hardly looked like someone battling cancer. His build was stocky but trim; he'd managed to start lifting weights again, he told the nurse, though not at full throttle. He had an easy smile and a gruff, hearty voice, and he was quick to respond to both doctor and nurse, eager to impress the professionals.

After a few more minutes spent talking about test results and the like, Druker, 56, knew the time had come. It was time for the biopsy, and he could see that his patient was nervous. "Are we done stalling?" he joked, trying to set his patient at ease. Eichner moved to the examination table, where he lay belly down, the thin white paper crumpling beneath him. A second nurse injected a dose of Ativan, a general anesthetic, into a vein in his arm, and Druker injected a shot of lidocaine, a local anesthetic, directly into his hip, where the needle would be inserted. A couple of feet away, a technician prepared the instruments, laying them out neatly on a metal table for Druker. Across the room behind a thin, floral-patterned curtain, another nurse had taken the seat by the computer, a pink streak jazzing up her blonde bob. She called out, "Gary, what's your name and date of birth?" "Gary Eichner, 10, 5, 69," he replied. The pain medication had started to relax him. "And what are you having done today?" she asked, following hospital protocol to confirm this information with the patient before every surgical procedure. "I'm having a bone marrow biopsy," he said. Her reply of "good job" was greeted by a deep laugh from the rapidly numbing Eichner.

The bright room, lit as much by the wall of windows as by the fluorescent ceiling lights, grew quiet. Small bursts of chatter among the two nurses, doctor, and lab technician at each step of the process arrived like waves onto a silent shore and quickly retreated. "You're going to feel a needle stick back here," Druker told his patient. "To avoid total discomfort, you have to breathe."

As a botanist bores into the side of a tree for a sample of the trunk's core, so does a doctor drill into the marrow. After making an incision in the skin, Druker, his lean runner's physique leaning over Eichner's back, drove a slender, bevel-edged, hollow cylinder, called a trocar, enclosing a sharp needle, or stylet, straight downward. He pushed the cylinder and needle through muscle and fat toward the iliac crest, the winglike bone that forms the top of the butterfly-shaped pelvis. Though fit and strong, Druker exerted all his strength twisting the handle at the top of the trocar back and forth to break through the bone. He leaned on Eichner for leverage, feeling the first tingles of perspiration as he continued to push downward.

It was this effort that had Eichner's nurse at his first biopsy calling for a hammer. Still stinging from that memory, Eichner was prepared for the worst, but Druker, his necktie tucked inside his checked shirt to avoid tickling his patient's flesh, worked quietly and efficiently. Any words he uttered were strictly to comfort his patient. Eichner's bones had softened now that he'd had the drug in him for six months, and it took Druker just a couple of minutes to get the needle inside Eichner's left hip bone.

Standing on the other side of the bed, monitoring the Ativan dose, a nurse asked Eichner every minute or so how he was doing. Leaving the needle sticking out vertically from the top of Eichner's buttocks, Druker unscrewed the stylet and inserted a syringe through the trocar. First was the "dry pull," an extraction of 1 or 2 milliliters of marrow. With this sample, a technician would check the size and shape of the cells, and the percentage of blasts. A second syringe drew about 10 milliliters of marrow; this one was the "wet pull," with heparin added to stop the blood cells from clotting. This sample would be sent to the genetics lab where, in a darkened room, technicians would count the number of cells housing the Philadelphia chromosome. If it were twenty out of twenty, the drug was not working—not yet, anyway. Anything less than twenty would satisfy Druker. Eichner was holding out hope for zero.

Druker completed the two pulls, handing over the samples of marrow, red like blood but shinier, to the technician, who verified the source of the specimen by the presence of spicules, white clumps interspersed throughout the body's marrow. Last was the bone biopsy. Druker inserted a new needle and dug farther into the hip bone. He removed the stylet and turned the trocar back and forth, shaking loose a bit of bone that remained stuck inside the cylinder. Druker pulled the stainless steel tube out of his patient's body and knocked the centimeter-long piece of bone into a small plastic canister.

Eichner's son, lanky and tall, with floppy hair and warm brown eyes, and his brother, slightly older than Eichner but with the same casual air, returned from the café soon after the procedure was complete, quickly displacing the awkwardness of the moment with mild teasing about Eichner's exposed bottom. It was the first time they'd met Druker, a moment that Eichner had been hoping for when he asked his son to join him on the trip. Eichner flipped over, woozy and slightly silly from the anesthetic. He buttoned his jeans and exchanged some parting words with his doctor.

The hole in Eichner's bone would take about two weeks to heal. The test results would be ready in about three weeks. In the best-case scenario—no cells containing the Philadelphia chromosome—this bone marrow biopsy would be Eichner's last, possibly for the rest of his life.

THREE WEEKS AFTER his biopsy, Druker's words were still bouncing around Eichner's head. "I will be happy if you're eighteen out of twenty." Eichner knew that he, too, should be happy with that, because that would mean he was responding to his medicine. It would mean he could keep going with the pills he'd grown accustomed to, but with no more experiments, no need to put his body through the wringer with a new drug. It would mean that he was safe. Just that much of a reduction would be enough to confirm that the cancer inside his bones was not going to kill him.

On March 6, 2012, Eichner received an e-mail telling him that his test results could be viewed on MyChart, the private online system that enables patients to access their medical records remotely. Before he had a chance to read them, Carolyn Blasdel, Eichner's main contact at Druker's clinic and the same nurse who'd questioned him about side effects, had left messages on his answering machine about the results. "She called me a couple of times, one after another, so I knew something was going on," said Eichner, who thought he heard a positive tone in her voice on the messages she'd left.

He'd seen some initial findings from his last visit. His white and red blood cell counts and some other information had been posted on MyChart just a few days after the biopsy. But he couldn't make much sense of the numbers. He thought they looked good, but what did he know? Plus, he didn't want to give himself false hope. The only number that mattered to him was how many out of those twenty sampled cells still contained the Philadelphia chromosome.

He called Blasdel from work, and she reviewed all the tests with him, explaining the meaning of words like FISH, karyotype, and Bcr/Abl. She opened up the cytogenetics report on her monitor and read the key phrases to him. "All twenty metaphase cells appeared normal male. All results were within the normal limits," the report stated. She translated the words into plain English: None of the analyzed cells had the Philadelphia chromosome. He was zero out of twenty. There was, Blasdel read to him, "no morphologic evidence of chronic myelogenous leukemia." And, his white blood cell counts had returned to normal.

Because the data on patients who've taken the medicine Eichner was taking are still evolving, clinicians now deliver prognoses to CML patients in five-year intervals. Based on the response rates so far, Blasdel explained to Eichner, he had a 99 percent chance of surviving the next five years.

Surrounded by coworkers and $20 million worth of electrical equipment at his job site, Eichner tried to contain his joy. "You just want to scream and go crazy," he said. As good as it felt to finally tell the news to someone at work, the best moment came a few hours later, "letting my son know when I got home that night that the next five years are good." Eichner would be there for his son for at least a few more years. He knew how lucky he was. "Ten years ago, I'd be almost dead right now," he said. "Now I have a five-year [outlook] that is almost a guarantee, as long as I take my medicine."

Three months later, Eichner got more good news. The latest tests showed that he was in complete molecular remission. The most probing analysis had revealed no evidence of abnormal cells. Although he couldn't be considered cured, since the disease would likely recur if he stopped taking the pills, the remission meant that Eichner was, for the foreseeable future, essentially cancer free.

PART 1

The Chromosome and the Disease

1959–1990

In 1959, the blood-based cancer known as chronic myeloid leukemia (CML) was universally fatal. Even when the disease was diagnosed in its earliest stages, most patients died within six years. The sole treatment, radiotherapy for the spleen, did little to improve the odds. Several decades later, treatments were still inadequate: Drugs gave many patients an extra few years, but eventually the malignancy became unmanageable.

CML was hardly exceptional. Even in the early 1980s, nearly every type of cancer remained stubbornly incurable. Those who treated it stumbled into a hopeless vortex, and almost all who had it faced an early death.

1

THE FIRST CLUE

David Hungerford could not believe what he was seeing. He hovered over a microscope, turning the wheels this way and that to ensure the best view. A small glass slide was illuminated from below. It held a single cell that had been expanded and then stopped in the middle of reproducing, its forty-six chromosomes on full display. He checked and rechecked, and was absolutely certain: One of the chromosomes was too short.

It was 1959, and the field of genetic research was almost nonexistent. The 1956 confirmation of the standard number of chromosomes housed in the human cell—forty-six, in twenty-three pairs, one set inherited from each parent—hinted at something impossible to grasp, a continent on a horizon too distant to see with the tools of the day. Even though James Watson and Francis Crick had made their famous discovery of the helical structure of DNA in 1953 and the genetic root of Down syndrome—an extra copy of one chromosome—had been found the same year, the search for connections between DNA and disease had only just begun. Around the world, laboratories were just starting to toy with the kind of technology needed to explore genetic matter. Genes were units of heredity, a way for traits to be passed on from one generation to the next, including deficiencies. But how disease could possibly be linked to DNA was entirely unknown. Phrases like "genetic mutation" or "chromosomal abnormality" were not part of the vernacular yet because there was no need for such language.

And so it was that David Hungerford, a young scientist hovering over a microscope, was stunned by what he was seeing through the lenses. This was a man who knew how chromosomes should look. Camera-equipped microscopes were hot laboratory commodities in the 1950s, and Hungerford, an avid photographer, had gotten a job working with one in a Philadelphia cancer research center. He spent countless hours looking at the starfish-shaped chromosomes of the drosophila fly, training his eyes to see the fine banding patterns within. He was one of a handful of people alive at the time who could have spotted an anomaly among a blurry, inky array of chromosomes.

So it may have been inevitable that he'd ended up working with Peter Nowell, a doctor also in his early thirties doing cancer research across town at the University of Pennsylvania. In 1956, Nowell had accidentally stumbled upon a new method for seeing chromosomes inside cells. He had been studying blood cells from leukemia patients, his work following the usual approach of the day: rinsing the cells and staining them with a bluish-purple dye.

Science had come a long way in its ability to peer inside cells, the basic structural units inside every living thing, since they were first spotted by microscope in 1665. That discovery led to others, which led to the creation of cell theory, the notion that all living things are made of cells, and that new cells are made when old cells divide. But the cutting-edge techniques for seeing the inner clockwork were still rudimentary, calling for the scientist to squash a drop of cells on a covered glass slide with the thumb in order to put pressure on the cells. The squash was supposed to burst the cell, spilling out its gene-filled middle. But the approach failed as often as it succeeded, leaving behind broken cell fragments that were useless to researchers. People were frustrated with the technique, which wasted precious time and resources.

One day Nowell took a shortcut around the usual scientific procedure. "Pete was in a hurry, as young men tend to be," Alice Hungerford, David's wife, would recount years later. Instead of following a more rigorous cleaning method, Nowell washed a sample of white blood cells under some tap water. He dropped the rinsed cells onto the slide and was amazed by what he saw through the microscope. The tap water, it turned out, was hypotonic—a low-pressure solution that caused the cells to swell, like a deflated raft being blown up with too much air.

With the cells ballooned like that, Nowell could see something else equally surprising. It turned out that a bean extract he'd applied to help clot the red blood cells (making them easier to remove from a sample) had also stimulated division in the white cells. Captured in the midst of dividing, the cells were at their most expanded. Because the tap water had further expanded the size of the cell, the chromosomes had more room to spread out and were suddenly easier to see and count. No one was looking at chromosomes this way. Nowell hadn't known it was possible. Then again, he knew nothing about genes and had little interest in genetics. But he kept the slide, figuring someone out there might be interested in taking a look.

The genetics community was small then, and the number of people in the Philadelphia area interested in genetic research could be counted on one hand. Hungerford heard about Nowell's slide. The two began working together. For years, Nowell prepared slides that Hungerford would study under the scope. They perfected the hypotonic solution, still used in molecular genetics today, and figured out how to air-dry slides to help the cells spread out even more. But they saw nothing noteworthy.

Then, in 1959, three years after they'd met, there it was: an abnormally small arm of a worm-shaped chromosome inside a cell of a person with CML. With the chromosomes splayed in the squashed cell, Hungerford could clearly see that one was too small. A piece of it was missing. They looked at blood samples from six other CML patients and found the same abnormality.

Stunned, Hungerford snapped the camera shutter. He would not live to see the significance of the picture he'd just taken. In 1959, the effect that a single photograph showing a single mutant chromosome would have on the lives of countless patients and on the future of cancer treatment was entirely unsuspected.

"Until we stumbled over this Philadelphia chromosome, there was really no evidence that cancer might be due to genetic change," Nowell, now 79, said decades later. This photograph would become the lasting portrayal of a moment when everything changed for cancer and medicine as a whole. It was the as-yet unrecognized starting point for the modern era of targeting cancer at its root cause.

2

THREE HUNDRED WORDS

At the time of their discovery, David Hungerford was spending about ten hours a day looking at fly chromosomes, and Peter Nowell had just returned to the University of Pennsylvania. Nowell had originally started working in the pathology lab there as a summer job in 1950. A cocky and charismatic med student, Nowell had felt certain that, given the chance, he could "solve this cancer problem" in a matter of months. But that summer he got married, and the Phillies were on their way to winning the pennant, distractions that, he says, delayed his plans to cure cancer.

But those few summer months were enough for Nowell to understand just how vast a territory he'd entered into as a cancer researcher. "I really knew very little about the specifics of things," he would say later. "In those days, it was true of pretty much everybody." He decided to take an internship for a year with a hematologist at a nearby hospital. It was there that he had his first serious education about cancers of the blood—how devastating and how complicated these diseases really were.

There were the leukemias that took over the white blood cells, with chronic versions that progressed slowly and acute versions that led to rapid destruction of the immune system. White blood cells, which fight infections, normally numbered 4,000 to 10,000 per microliter of blood. Leukemia patients typically had counts in the hundreds of thousands per microliter. The lymphomas, Nowell learned, poisoned the lymph, another infection-fighting part of the immune system concentrated primarily in bean-shaped nodes throughout the body. Lymphoma could pass from one node to another, like a fungus spreading through a forest. Multiple myeloma targeted the plasma—the yellow-colored liquid that holds red blood cells, white blood cells, and platelets in suspension as they course throughout the body—hardening the fluid into tumors inside the marrow and softening bones.

These were the so-called liquid cancers, or hematologic malignancies. For cancer researchers, they were often easier to study because of their accessibility. It was much simpler to draw fluid out from a vein than to cut a patient open to excavate a solid tumor buried deep inside the body. But being able to get at the cancer hadn't led to more significant advances in treatment. When Nowell was in medical school, most types of liquid cancers were still incurable.

On rounds, he saw the victims of these harsh diseases. They brought to life the horrors of cancer more than any squashed cell ever had. A young person getting his first palpable glimpse at death, Nowell saw how shallow he'd been when he first arrived at Penn. Humbled, he realized that cancer was a beast the world had been wrestling for centuries, and against which few meaningful strides had been made.

Then, just when he was feeling a surge of dedication to the long haul of cancer research rise up in him, Nowell was drafted into the military. He was sent to San Francisco to work at the US Radiology Defense Laboratory, where he was assigned to a team studying the potential effects of radiation. The government wanted to know the possible dangers associated with the fallout from nuclear testing in the Pacific. The risks to people included diminished numbers of red and white blood cells circulating in the body in the short term and, in the long term, leukemia and other malignancies. Again, the horrors of cancer were made ever more apparent to him—this time, even more so as he witnessed the man-made devastation.

In 1956, Nowell returned to Penn, as determined as ever to solve this cancer problem.

Hungerford, on the other hand, had no desire to cure cancer. It just wasn't his way as a scientist. He had taken the more scholarly PhD route, and the driving force behind all of his work was a love of observation—to look, to record what he saw, and to share those findings with anyone else who might be interested. Compared with Nowell's passion, Hungerford's approach could seem cold and distant to their colleagues. But Hungerford was happy to record their observations for the simple reason that observations should be recorded. "He just liked to look through the microscope and see the thing," said Alice. He felt no ownership of his ideas, and he had no need for recognition. He just wanted to do the work of science; that was his role in the world. It was what made him feel alive.

Nowell and Hungerford's discovery of "the minute chromosome" was published in 1960. The report consists of three brief paragraphs in a scientific journal, without even the typical list of references that scientific papers have, set indiscriminately among a few other reports of the month. "It's three hundred words," said Emil Freireich, a leukemia doctor responsible for many major therapeutic advances, and a towering figure in the world of cancer medicine. "And it revolutionized everything."

When Nowell and Hungerford published their third scientific paper documenting the truncated chromosome in a large number of patients, with reports from groups at universities around the world confirming the phenomenon, the minute chromosome was renamed the "Philadelphia chromosome" in recognition of the city where it had been discovered.

After scientists across the world found the abnormal chromosome in their own CML cell samples, many set to work on finding other such mutants. At first, researchers thought that this chromosome was the first drop in what would soon become a waterfall of genetic mutations linked to cancer, and, they hoped, some meaningful advancement for cancer treatment. In scientific journals, the chromosome was referred to as Ph^1, an abbreviation that left space for other mutation discoveries to come—Ph^2, Ph^3, and so on—with researchers in other cities then following suit. But further cancer-linked mutations proved elusive. No others were found, in Philadelphia or anywhere else. Ph^1, as it is still often called, was found in a small percentage of samples from patients with other types of leukemia, acute lymphoblastic leukemia and acute myeloid leukemia (AML), but the link was not nearly as strong as that seen in CML. There was a brief stir over an abnormality spotted by some New Zealand researchers—the "Christchurch chromosome," people called it—but that soon turned out to be a false alarm. Whatever mutations were found appeared much more rarely than the Philadelphia chromosome did in CML. Those tenuous links hardly seemed the stuff of cancer cures.

And so enthusiasm over the Philadelphia chromosome waned, mainly because no one knew what to do with the information. "In the early years, the medical community did not care about human chromosomes," recalled Alice Hungerford, who met David when she took a job in his lab. It was like seeing a bright spot in the night sky with no knowledge of planets and solar systems. Despite the obvious connection between CML and the Philadelphia chromosome, there was very little suspicion of a causative link between genetic abnormalities and cancer. There was no technology to look any further into the mutation. In fact, it wasn't even called a mutation; it was considered a deletion. Nowell and Hungerford resisted the notion that the piece of chromosome was completely gone from the cell. They knew that such a deletion would likely be lethal. But they had no explanation for what else could have happened. A piece of genetic material had vanished. Why had it disappeared? Did the change somehow cause leukemia, or did leukemia somehow cause the change?

These were questions for another decade. Knowing the standard number of chromosomes had enabled geneticists to create a universal number language. But the view afforded by the technology at the time was so coarse that at first Nowell and Hungerford couldn't even tell which chromosome the abnormality was located on. Eventually it became clear that the deletion was from one of the two copies of chromosome 22, but that was still an incomplete description. Chromosome 22 looked an awful lot like chromosome 21 and sometimes even the Y chromosome present in males. Later, methods for staining specific chromosomes would allow for a much more discriminating study. But in 1960, these techniques were unknown. Whatever questions scientists had about the Philadelphia chromosome, there weren't any that could be answered.

NOWELL AND HUNGERFORD's collaboration also reached a standstill. It was as if they had come together just to find the Philadelphia chromosome, and now, having done so, needed to move on. Nowell continued to pursue cancer research, and would ultimately spend his entire career in the same laboratory. His early success garnered him a rare lifetime government research grant. The money enabled him to pursue cancer research without the pressure to churn out publications or complete grant applications every few years, a highly limiting factor on lab research today. "I had it easy," Nowell recalled toward the end of his years at Penn, his hair as white as his lab coat. "As my wife used to say, I just assumed there was a closet with green pieces of paper in it." Nowell's grant left him free to continue research without worry about making discoveries.

Although he never again struck gold with a serendipitous discovery, Nowell contributed to important theories about how tumors evolve. He was an early adopter of the notion that tumors accrue mutations over time, a key component of modern anticancer drug development. As he put it, cancer works like a tree. A branch off the trunk is the first mutation, and every subsequent twig represents further changes to the DNA. In the end, a cell that started off just a bit different from normal accrues multiple oddities, each one enabling it to better survive in the body, and each one a potential target for a new drug. This phenomenon is at the heart of current cancer research, as scientists sift through the dozens, sometimes hundreds, of genetic abnormalities for the ones that are advancing the deadly cells.

Hungerford's life took quite a different turn. In 1971, he was diagnosed with multiple sclerosis. Not wanting to suffer the pity of his peers, Hungerford kept his disease a secret, telling only Nowell, who had become his friend and now confidant. When Hungerford's treatments and condition diminished his productivity, colleagues and grant reviewers assumed he was either lazy or untalented. His funding gradually decreased, and eventually his lab at Fox Chase Cancer Center, where he'd worked since before meeting Nowell, was shut down. Devastated, Hungerford never stepped behind the microscope again. "He did not pick up a scientific journal after that," says Alice. "It broke his heart." A longtime smoker, Hungerford died of lung cancer in 1993 at the age of 66.

3

INVESTIGATING A CHICKEN VIRUS

Like the seed that withstands a drought only to bud with the first touch of rain, further investigation of the Philadelphia chromosome would lay dormant until the early 1970s. The blossoming of the research was largely driven by technology, and in the early 1960s, the tools needed to parse the meaning of this chromosome were nonexistent. They were still more than a decade away.

In the meantime, a second thread of research was ongoing, one that would eventually connect to the Philadelphia chromosome, but in a wholly unexpected way. This was a story not of genes but of viruses.

Almost hidden away on New York's Upper East Side, Rockefeller University, founded in 1901, thrums with a quiet, brainy intensity that has led to countless scientific discoveries. Twenty-four Nobel Prize winners have graduated or conducted research there. Lab work at Rockefeller has led to breakthroughs in understanding hepatitis B, obesity, diabetes, cancer, skin diseases, infectious diseases, and other ailments and puzzles. In its first decade, when it was still called the Rockefeller Institute for Medical Research, a man named Peyton Rous studied lymphocytes, white blood cells that flow throughout the body and gather in small clusters, known as lymph nodes, in the underarms and other regions of the body.

Rous, born in 1879, and his two siblings were raised by his single mother in Baltimore, Maryland. After her husband died, she'd resisted the temptation to return to her home state of Texas, where family could help raise the children. Instead, she stayed in Baltimore, believing that the city would afford her children the best possible education. It was not school but the wildflowers he encountered on his frequent walks through the woods near his home that first bent Rous toward science. A month-by-month chronicle of the flowers he observed on those walks, printed in the *Baltimore Sun*, was his first published paper.

After earning his medical degree at Johns Hopkins, Rous decided to focus on medical research instead of a medical practice. He'd worked as a pathologist at the University of Michigan from 1906 to 1908, work that included a 1907 stint in Dresden, Germany, where, he would later write, there was "no hint of war in the air." He joined Rockefeller as a laboratory researcher in 1909. In September of that year, after a few scant weeks on the job, Rous received a visit from a Long Island farmer, her arthritic hands carrying a Barred Plymouth Rock hen, a common domesticated chicken. A large tumor was sticking out of its breast, poking through its striped feathers. The chicken, "a strong, young hen," Rous would say in the 1910 paper he published about this work, was about fifteen months old, the tumor about two months. Its owner wanted Rous to remove the hardened malignancy.

Rous agreed, and a few days later Rockefeller Institute had its first-ever chicken surgery. On the first day of October, Rous numbed the hen with ether and sliced into its belly. He removed most of the irregular, globular mass, its yellowish-pink tissue crumbling under Rous's knife. The operation was not successful: The hen died a month later from remnants of the tumor spreading in the tissues around its abdomen. But the excised tumor created a lasting legacy.

Not wanting to waste the opportunity, Rous decided to take a closer look at the excised tumor. He removed a portion and ground it up finely, then forced the crushed mass through an extremely fine filter, one that would prevent the passage of chicken matter or bacteria to the other side. The theory was that the resulting filtrate would contain only the elements that were directly relevant to the growth of cancer. When the extract was injected into other fowl, the animals developed the same knobby tumor within weeks of inoculation. The cancer was transmissible. There was only one explanation for this transmissible chicken cancer: the underlying cause had to be a virus.

Rous was not the first to show a connection between viruses and cancer. An Italian scientist in the 1840s had observed that nuns in Verona got cervical cancer only rarely, whereas married women experienced it more frequently. The culprit—human papilloma virus—would not be identified until 1983, but it was clear then that some contagious element had to be at play. Just a year before Rous was presented with his famous chicken, two Danish scientists had shown that a certain type of bird leukemia had the same infectious quality, but the discovery was largely ignored because that leukemia was not recognized as a malignancy at the time. A type of lung cancer seen in sheep was known to be transmissible in the 1800s, and an infectious anemia virus had been found in horses in the early 1900s. But none of these had the impact of what would come to be called the Rous sarcoma virus, or RSV. His was the first concrete demonstration that cancer could be triggered by an infection.

Yet it wasn't until the 1960s that RSV would really enter the cancer research limelight. At the time, interest in viruses as vectors for passing cancer was on the rise for two reasons. First, there was mounting suspicion—based on the sarcomas like the one Rous had seen and other animal cancers—that viruses cause human cancer. If animals got cancer this way, why not people? Medical research was desperate for any glimmer of understanding about the genesis of cancer, and what with their invisibility and skills of invasion, viruses seemed like a feasible answer.

But equally as important, viruses were proving a useful tool for studying how cancer transforms healthy cells into the out-of-control masses that eventually kill the people in which they grow. With their ability to cause cancer in lab animals such as mice, viruses provided a vehicle for observing cancer in a controlled research environment. At a time when people knew nothing about how or why cancer occurred, any possible inroad was worthy of attention. "The level of ignorance was such that anything was valued as a potential clue," said Stephen Goff, a Columbia University virologist whose research, begun in the 1970s, was instrumental in moving the field forward from these early days.

Throughout the history of cancer research, knowledge has advanced in step with technology. In the 1950s, a virologist named Howard Temin set out to find a new and improved way to study how viruses—RSV, specifically—cause cancer. After all, Temin reasoned, here was a virus that was known to transmit cancer. If he had a way to watch that happen, to witness the event outside of chickens and mice, in the cold light of a petri dish, then he might be able to unravel the minute mechanisms at play. In 1958, Temin and his colleague, Harry Rubin, succeeded.

The technique, called a focus assay, used a background of normal cells to bring cancer cells into stark relief. After exposing a culture of cells to a cancer-causing virus, the cells are smeared onto a dish and left to replicate. Cells that have turned cancerous will replicate faster, piling up in clusters that look strikingly different from normal cells. Now, they could quantitatively measure the transformation—how fast it happened, how severely, how much virus was required to launch the change. With the focus assay, Temin made the startling finding that a single particle, or virion, of RSV was enough to turn a cell cancerous.

Having a way to watch the transformation of virus-infected cells in turn transformed the entire field of cancer research. "Every day that a new technique came on line meant whole new things were possible to do," Goff said. The creation of that assay was like introducing push-button dialing into a rotary world. It sped up cancer research enormously and brought Temin worldwide recognition. It wasn't long before the assay led to the first shocking observation: RSV was a virus made up of RNA, not DNA.

Considering all the havoc they wreak, it's surprising how tiny and simple viruses can be. After all, they aren't even considered to be living things on their own because they can't reproduce outside of a host. Viruses come in two basic varieties: DNA and RNA. Both types consist of genetic material bundled inside a coating made of protein, and sometimes also some fat molecules. They typically contain very few genes, often just four or five. In the early days of virology, DNA viruses were thought to be the ones most relevant to cancer. But thanks in large part to Temin's focus assay, it gradually became clear that RNA viruses could also transform cells from healthy to cancerous.

The presence of RNA, rather than DNA, in the cancer-causing RSV was a puzzle. Every scientist of the day knew that the DNA synthesis at the heart of cell division—the process by which cells multiply, and which occurs continuously throughout our lives—followed a well-worn path: DNA is translated into RNA, and RNA is translated into proteins. In light of the variety of cells in a single body, let alone the diversity of life, DNA replication is strangely simple. The entire genetic code is spelled out in patterns of just four nucleotides, symbolized by the letters A, T, G, and C (for adenine, thymine, guanine, and cytosine), each paired with another (A with T and G with C) in complementary strands held together in a spiraling ladder. The major visible difference between RNA and DNA is that the former contains uracil instead of thymine. As the DNA helix unwinds during cell reproduction, a complementary strand of RNA translates its pattern of base pairs into proteins. DNA is the instruction, RNA delivers the message, and proteins carry out the instruction.

There were DNA viruses known to cause cancer in animals, and how that happened was mystery enough. But RNA viruses were another story. Because it's the go-between and not the final form, the RNA inside such viruses should not be able to cause permanent damage to the human genome. Poliovirus, comprised of RNA, was a terrible disease but it didn't become integrated into an individual's DNA; when the virus was gone, it was gone. The same was true for influenza, another RNA virus. When we catch the flu, it doesn't become part of our genetic makeup and leave us passed out for the rest of our lives because our DNA now contains flu virus genes. Eventually, the virus is eradicated from the body, leaving our genome intact.

Scientists of the day knew that cancer was somehow connected to change at the genetic level, but the assertion was based more on logic than evidence. The thinking went that if a disease proliferated across generations of cells—that is, the disease is reproduced in each new cell, indicating that the cells were irrevocably changed, as they are in cancer—that some permanent genetic alteration had to be at play. But the idea of an RNA-containing virus causing such a change seemed preposterous.

With nothing to back up his assertion, Temin pronounced that RNA viruses made DNA, and that this DNA would be present in the tumor genome. This thinking went so far against the grain—everyone knew that DNA made RNA, not the other way around—that it prompted outright ridicule of Temin. "[He] was absolutely considered nuts for many years," said Goff, who, though a boy at the time, would soon enough be immersing himself in RNA virology.

Among the few who did not consider Temin to be insane was David Baltimore, a rising-star virologist at MIT, who joined Temin in his effort to find the secret ingredient that enabled RNA to convert to DNA. Temin had hypothesized that the virus contained an enzyme—a protein that often operates as an assistant to various cellular processes, helping to speed them up and run smoothly—that enabled the reverse process of RNA's conversion into DNA. And, in 1970, that was exactly what they both found. They named the enzyme "reverse transcriptase," and RNA viruses were later renamed "retroviruses" because of their retrograde approach to reproduction. RNA viruses that cause cancer came to be called "oncogenic retroviruses." Baltimore and Temin shared the Nobel Prize in Physiology or Medicine in 1975 (Temin was also recognized for his focus assay with Renato Dulbecco, a pioneering virologist at the Salk Institute who had been Temin's mentor).

Virologists around the world continued to study RSV along with other so-called transforming viruses with the aim of understanding how a viral infection could cause cancer. A scientist named Hidesaburo Hanafusa, then at the University of California—Berkeley, made a crucial finding when he figured out, through a series of painstaking experiments, that some transforming viruses were actually mixtures of two viruses (though Rous's virus was not this type). When detangled, one of the viruses would induce cancer transformation but couldn't replicate, and the other could replicate but wouldn't induce cancer transformation. Hanafusa knew that the only difference between the two viral strands was a single gene. If one strand lost the ability to cause cancer when it was separated from the other strand, then clearly there had to be some single gene responsible for triggering cancer.

The next breakthrough came from the study of mutant versions of RSV that were sensitive to changes in temperature. In each version, a different gene in the viral genome was rendered defunct by a change in temperature: The gene was induced to operate normally at a lower temperature but not at a higher temperature. By switching off individual segments of the RSV RNA, scientists could discern their various functions (although RSVs hold their genetic information in the form of RNA and not DNA, the information still encodes distinct genes). When the temperature was raised, what was different about the virus? Those differences helped pinpoint the nature of the temperature-sensitive gene. It was like trying to determine the window through which an intruder had entered. Each temperature-sensitive mutant provided clues about which one may have been broken into.

In 1970, G. Steven Martin, working in the lab of Harry Rubin—Temin's old focus assay colleague—isolated a mutant strain of RSV that induced cancer transformation at 35° Celsius but not at 41° Celsius. The virus did continue to replicate at the higher temperature. Martin now knew that the portion of RNA in the virus that had been made temperature sensitive in that strain was necessary for tumor formation and tumor formation alone. That was where the break-in had occurred.

The gene became known as *src,* pronounced "sark," for the sarcoma it caused in chickens like the hen that Peyton Rous had examined sixty years earlier. "It was the most exciting idea at the time in cancer research," said Ray Erikson, a Wisconsin farm boy who was about to enter the RSV fray. "Here you had apparently a single gene in a virus that, when it was turned on, could transform a cell population and cause tumors in chickens." Very soon, studies of *src* were going to shake cancer research to its foundations and lead to sweeping changes in the research and treatment of this disease, beginning with CML. But first, someone had to figure out what, exactly, *src* did. How did this one tiny strand of DNA turn perfectly healthy cells into chicken-killing cancer? What was the exact chain of events inside the cell that led to tumor formation, and how was this gene involved?

Finding out how *src* could do such a thing meant figuring out the protein it encoded. The science at the time said that every gene coded for a specific product in the cell, a mechanism integral to survival. What was *src*'s product? "I regarded that question as probably the most important question that I could work on in my small laboratory at the University of Colorado," said Erikson, who had left his family farm to pursue a career in science. Erikson and the members of his Denver lab set to work to try to find that single protein against a background of hundreds of thousands of other gene products in the host cell. It was the needle in a haystack taken to an outlandish extreme. Diving in to find that needle set in motion a second trajectory of research—the first being the Philadelphia chromosome—that would prove crucial to uncovering one of cancer's most deadly schemes.

4

RIGHT NUMBER, WRONG PLACE

As Erikson was getting under way with his search, reports from England about new techniques for coloring chromosomes surfaced. Up until 1969, Giemsa staining, a process using a dye originally created to test cells for the presence of malaria and other parasites, was the only available method. It was effective—chromosomes, derived from the Greek for *colored body*, had been named as such because they so readily soaked up the dye, enabling the otherwise clear structures to be seen. But Giemsa staining could only turn chromosomes into a single uniform color, which, as Nowell and Hungerford found out, was useful but had serious limitations. Now, there was chatter about new "banding" techniques that lit up chromosomes in a completely new way.

Chromosomes look like pairs of stubby worms belted together at their middles, or, in more graceful depictions, headless dancers. They float in a sea of jelly-like plasma within the center of every cell in the body. Composed entirely of genetic information, on a closer look each of these inky blobs appears as a tightly curled squiggle, a single DNA molecule and its associated proteins, compressed into a spiraling ladder. In humans, these tightly packed helixes contain about 20,500 genes altogether. (The Human Genome Project and the subsequent Encyclopedia of DNA Elements [ENCODE] pilot revealed that just 1.5 percent of the human genome encodes proteins and that, though the vast majority of the remaining genes are doing something, their exact biochemical function is still largely unknown.)

In the early years of genetic research, scientists often focused on other species to learn how chromosomes work. Methods for using blood or bone marrow hadn't yet been developed, and finding volunteers willing to donate tissues that were most amenable to genetic research, such as the gonads, was not easy. The first key discoveries were made by observing sea urchins and horse roundworm. In the late nineteenth and early twentieth centuries, the German scientist Theodor Boveri proved three essential facts about chromosomes: that they are the carriers of heredity, that each chromosome contains different genetic information, and that every developing egg receives a full set of chromosomes from each parent—to make an embryo, the same story of life must be told twice. These three facts are the tripod supporting the entire field of genetics.

Boveri's experiments also led him to believe that cancer is caused by genetic abnormalities, a theory that stands at the center of modern cancer research. In 2006, the National Cancer Institute launched the Cancer Genome Atlas, one of many worldwide efforts to catalog the genomes of all different types of cancer. Just as the Human Genome Project spelled out the nucleotide sequences for the 20,000-plus genes that make up human DNA, the atlas maps the sequences of A, C, T, and G in cancer, where the slightest changes in those patterns can alter the instruction issued by a gene. Switch an A for a T, and a gene that encoded a protein that helped program a cell to die at the right time could cease to operate, leaving the cell essentially immortal, a trait of many malignancies. Alterations in nucleotide sequences can be inconsequential to a cell, or they may disappear just as quickly as they arrived. But some are deadly. Spotting the crucial changes in gene sequences, the ones that are responsible for triggering tumor formation, growth, and survival, is a central goal of cancer research today.

Boveri was onto this trail in 1914, though the idea would garner little, if any, traction among his peers. Boveri's idea was like a visitor from the future that no one could understand. As little clue as Nowell and Hungerford had about why CML cells had a strangely small chromosome, the scientists of Boveri's day had even less of a framework for understanding that cancer could be linked to genes. They also had no way to compare the chromosomes of cancer cells with those of normal cells. It would be decades before such a framework would be built, as the field of cytogenetics, the study of the connection between genes and disease, gradually came to life. First and foremost was figuring out how many chromosomes humans have. Without this number, all other inquiry was useless.

Early in the twentieth century, the effort to get the right count of human chromosomes was wild and haphazard. The numbers ranged from 15 to 115, with each researcher clamoring for his declared estimate to be accepted as correct. Finally, in 1921, Theophilus Painter, a world-famous zoologist at the University of Texas, pronounced what he considered to be the final word on the subject: Humans have forty-eight chromosomes.

It wasn't a bad estimate considering the material he was working with. Painter had been counting the chromosomes inside the sperm of testicles taken from castrated mental patients. But the chromosomes inside of sperm cells are unreliable in their appearance. Some chromosomes appear as doubles, and their messy arrangement makes them hard to keep track of. It could also be that the samples he was working with were abnormal, considering they had come from people who were seriously ill. But regardless of how close to correct it was, forty-eight was still wrong.

The mistake would not have been so bad if scientists to follow did not feel beholden to confirm Painter's finding instead of just reporting what they saw. Because Painter was such a renowned scientist and because he'd been so certain he was right, forty-eight became the dogma, and no one wanted to contradict it. Plus, technology was so rudimentary that whenever scientists did come up with an alternative number, they had little confidence in its accuracy. The loudest answer became the accepted answer, delaying genetics research by more than thirty years.

It wasn't until 1955 that the mistake was corrected. It was Joe Hin Tjio, an Indonesian scientist working in a Swedish laboratory, who figured it out. Tjio had fled his home country following his imprisonment by the Japanese during World War II. He'd been tortured for offering medical help to his fellow inmates and had kept himself sane by knitting clothes and refusing to give in to despair. Before the war, Tjio had been making headway with his efforts to breed a disease-resistant potato, and upon his release the Spanish government offered him a position with its own plant-improvement project. During holidays and summer breaks, Tjio went to Sweden to work with Albert Levan, a famed geneticist who was doing pioneering work with cytogenetics.

In the winter of 1955, Tjio and Levan were studying lung tissue from human embryos. One late December night, his enthusiasm unaffected by the frigid weather and deep snowdrifts, Tjio trudged to the lab to prepare some slides for the microscope. Levan had discovered that a few drops of colchicine, a plant-derived toxin, caused mammal cells to halt right in the middle of dividing, a feature that had already been seen in plant cells. It was like stopping the music and telling the dancing children to freeze, providing the ideal moment to make a head count. Tjio dropped the colchicine onto the lung tissue cells, let the mixture sit for a few hours, and made a slide squash. He put the slide under the microscope, switched on the light, and focused the two small lenses on his target.

Tjio had no intention of counting chromosomes that night, but he couldn't help himself. The colchicine technique had enabled Tjio to view the chromosomes more accurately than any scientist before him, including Painter. And there was no denying what he saw: forty-six chromosomes. Tjio made the bold move of reporting his results, risking the judgment and ridicule of peers who had cleaved to Painter's count for so long. But as Tjio's colleagues around the world examined their own cell samples anew, the only response was confirmation.

FIFTEEN YEARS LATER, Janet Rowley, a geneticist at the University of Chicago who knew well the peculiar history of chromosome research and was already an avid researcher of the connection between genes and cancer, heard about the new staining methods that had been created in England. Right away, she knew she had to learn them. Rowley was a scientific genius who'd graduated from college at age 19 and was the only woman in her class at medical school. Fortunately, 1971 was a sabbatical year for her, and, never wanting to be too far from a microscope, she decided to spend it with her husband, a scientist also on leave that year, in Oxford learning about the rumored new chromosome banding techniques.

The method was tricky but entirely rewarding. Before applying the Giemsa stain, the cell samples needed to be pretreated with quinacrine mustard, a powerful fluorescent dye. When viewed under a fluorescent microscope, the sample was no longer monochromatic. A pattern of yellow-green stripes, varying in brightness, ran up and down each chromosome. It was as if, having always watched movies with the lights on, someone finally thought of turning them off. It was a sight to behold.

Now geneticists could learn the ins and outs of the forty-six chromosomes—twenty-three pairs, one set inherited from each parent—in unprecedented detail. The patterns of stripes within each chromosome arm differed, and after just a few months of practice, a person could know instantly which chromosome he or she was looking at. With banding, each chromosome became a fully realized picture, poised to reveal the inner workings of the human genome and the mysteries of genetic diseases. The looming challenge was to understand what part of the human story each chromosomal picture was telling.

As always, Rowley's sharp intelligence served her well. She had the method perfected by the time she returned to Chicago. She eagerly resumed her work, armed with her new skills. She'd been studying a condition characterized by unexplained anemia and bone marrow abnormalities that is often a prelude to leukemia. Before her sabbatical, Rowley had found a genetic abnormality in people who developed this so-called preleukemia. She had also been looking at cells from patients with CML and had seen additional genetic abnormalities beyond the Philadelphia chromosome in samples from patients who'd entered the blast crisis stage, the culminating phase of the disease.

Using the new banding techniques, Rowley wanted to see if the assorted abnormalities she'd spotted among CML patients were in any way consistent. Did each person experience the same progression of mutations—the same branching off of twigs, as Nowell might have explained it—or did further mutations after the Philadelphia chromosome occur willy-nilly, with no predictable sequence?

Every day, Rowley would examine slides under the fluorescent microscope and take pictures with the attached camera. She could have the 35-mm film developed at any photo store, right alongside her family photos. Night after night she sat at her dining room table, cutting out photographs of chromosomes to study the patterns, her four children wondering who was paying her to play with paper dolls.

All of her samples had an abnormally small chromosome 22, the Philadelphia chromosome. She could see that one of the two copies of chromosome 22 in CML cells was missing some portion that was present on chromosome 22 in normal cells. But then, in 1972, she noticed something else. The samples also had an abnormal chromosome 9. With the banding patterns now revealed, she could see that the stripes among CML patients were different from those among healthy individuals. In cells from CML patients, chromosome 9 was longer than in normal cells. The difference wasn't much; without the banding techniques, she might never have spotted it. But with the fluorescent dyes working like invisible ink on a blank page, a new message had emerged.

She checked samples of CML that had been obtained earlier in the course of the disease. The alteration in chromosome 9 was already there. Even when none of the other abnormalities she'd documented were present, chromosomes 9 and 22 were altered. The chromosome 9 abnormality, appearing at exactly the same time as the shortened 22 mutation known as the Philadelphia chromosome, had been there all along. What's more, Rowley could see that the banding patterns missing from chromosome 22 mirrored the addition she was seeing on chromosome 9. There was no way for her to be absolutely certain, and yet she knew it had to be true. The piece that was missing from 22 hadn't disappeared at all. The vanished stub of chromosome 22 had migrated to chromosome 9. And banding that she could see on a normal chromosome 9 but that was absent in samples from CML patients was now on chromosome 22. The Philadelphia chromosome wasn't a deletion, as so many scientists had thought. It was, in the terminology of geneticists, a reciprocal translocation. Genetic material from two chromosomes had switched places.

The idea of a translocation, rather than a deletion, seemed crazy to many scientists at the time. Translocations had been spotted, but rarely in cancer. Just as Painter's erroneous chromosome count had become engrained, so had the notion that the Philadelphia chromosome was a deletion. This time around, though, acceptance came much more quickly. Rowley had already reported another translocation in acute myeloid leukemia (between chromosomes 8 and 21), and her peers soon came around. When they banded the chromosomes in their own samples of CML cells, they could see that she was right.

Yet the paradigm that chromosome abnormalities were of no consequence still prevailed. "They were dismissed as wholly unimportant, the result of genetic instability," Rowley would explain years later, her broad face framed by her loosely bunned hair. Alterations in chromosomes—additions, deletions, translocations—were clearly indicators for various diseases, but they were thought to be a neutral presence, a curious sideshow. As far as nearly everyone involved in the research and treatment of cancer was concerned, genetic abnormalities had nothing to do with the causes of this disease. Even as the evidence continued to mount, the thinking that had brushed aside Theodor Boveri's assertion sixty years before still held its grip.

But when Rowley found a third translocation in chromosomes 15 and 17 in cells from patients with acute promyelocytic leukemia, a variety of AML, she knew that this thinking was wrong. The existence of three independent translocations in three separate cancers was undeniable evidence that there was some essential connection between these translocations and the development of cancer. Despite the lack of professional support, she clung to the notion that somehow these translocations were causing leukemia. In particular, the one involving the Philadelphia chromosome—referred to as t(9;22), the t for *translocation* and the two numbers indicating the relevant chromosomes—which appeared even before the cancer had taken full effect, had to be triggering this dangerously unregulated white blood cell growth. But how?

5

THE SURPRISING SOURCE OF THE CHICKEN CANCER GENE

Beginning in 1970, just as Rowley was preparing for the sabbatical year that would enable her to learn about chromosome banding, Ray Erikson's unrelated search for Src, the protein product coded for by the *src* gene, was getting under way.* Erikson knew that this gene was the one responsible for the ability of the Rous sarcoma virus to cause cancer. The temperature-sensitive mutant experiments had revealed that much. But for the finding to be of any use, scientists had to learn what that gene did to turn healthy cells malignant.

Years passed with nothing to show for their work. As Erikson and the young postdocs he mentored watched one Denver winter after another come and go, the experiments yielded almost nothing. In the meantime, his search was the butt of good-spirited jokes among his peers. "A lot of my colleagues in the field would say, 'Ray Erikson only wants to cure cancer in chickens,'" Erikson recalled. After all, it was a hen from which Rous had surgically removed a tumor decades earlier, and RSV didn't occur in people. Still, Erikson—and many others—cleaved to the idea that the mechanisms underlying the transformation from normal to malignant cells would have similar qualities across

* Following scientific convention, genes in this book are denoted in lowercase italics (hence *src*), and proteins are capitalized and in plain font (hence Src).

many species. "We weren't concerned about the fact that we were working with chickens," said Erikson.

CURIOSITY ABOUT *src* extended well beyond Erikson's lab. While he tackled the question of the gene's protein product, others turned their attention to the gene itself. Where had *src* come from? How had a virus that, everyone knew, could replicate just fine without the *src* gene come to contain this cancer-causing sequence? Knowing how it had gotten into the chicken could provide important clues about how cancer worked.

Around the time that Erikson had begun his Src protein search, a theory called the oncogene hypothesis (the prefix "onco" comes from the Greek word for "bulk" or "mass") was gaining traction. Two scientists at the NIH had put forward the idea that cancer was caused by oncogenes—that is, genes specifically programmed to trigger this deadly disease—that had been deposited into the genome of vertebrates by viruses eons ago. These oncogenes, the theory went, would remain latent unless they were activated by some environmental carcinogen.

Over at the University of California–San Francisco, J. Michael Bishop, who ran a lab researching cancer genetics, found the theory dubious. "[It] didn't make a lot of evolutionary sense," Bishop explains. "A gene that has no purpose in a cell does not last very long." Bishop figured that such a gene would not have remained in the DNA of the species in question because the gene had no role in survival. But the question remained: Where had the *src* gene come from? How had it become part of the viral genome? By altering the DNA, researchers had created variations of RSV that did not induce cancer. If a genetically mutated strain of RSV failed to transform cells from healthy to cancerous, then clearly there had to be a gene responsible for transformation. If the virus could lose the gene and still continue to replicate, then that cancer-inducing gene wasn't inherent to the virus's survival. Inducing cancer was a gratuitous trait that, though it may have somehow enhanced the virus, was not inextricably tied to its existence. Replicating was what viruses did. A feature unrelated to replication—such as causing cancer—had to be extra, added to the virus after it evolved. So where had the gene come from?

Why was it part of the RSV genome, and how had it gotten there? Answering these questions set in motion a third crucial pathway of research. First was unraveling the Philadelphia chromosome, then searching for the Src protein, and now tracing the *src* gene back to its source.

Bishop knew that the gene had to have come from somewhere outside of the original viral genome. The work by Hanafusa, Martin, and others more than a decade earlier had also helped lay this foundation of understanding. But the question of where the gene had come from remained a stubborn riddle. The idea that a cancer-causing gene could *originate* in the animal that got the cancer simply did not exist. Like the invention of flight or the first notion of a round Earth, the concept would have seemed outlandish at the time.

For Bishop, as unlikely as the oncogene hypothesis seemed, it was as good a starting place as any. So he and Harold Varmus, a scientist who'd joined Bishop's lab to learn about viruses, began searching for the *src* gene in chicken DNA. Today, this work would be simple: The viral genes would be cloned quickly, perhaps even by an outside company, and finding whether there was a match in the chicken genome would be easy. But Bishop and Varmus were sending out their search party in the era before recombinant DNA, and creating effective foot soldiers and flashlights was just as difficult as finding a matching sequence in the vast genetic forest.

First, Bishop and Varmus took the RNA from the RSV genome and made a complementary strand of DNA, just as the virus does when it replicates inside the host. They rendered that DNA radioactive. Then, they took RNA from a mutant version of RSV that did not contain *src*. They zipped together the mutant RNA with the radioactive DNA. It was like pasting up two strips of wallpaper so that the pattern lined up correctly at each edge.

Because the radioactive DNA strand and the RNA strand were complementary, Bishop and Varmus knew they would match up just as DNA and RNA did in normal cell division. But because the RNA strand lacked the *src* gene, the code for the *src* gene on the radioactive DNA was left out of that match-up, leaving Bishop and Varmus with a piece of radioactive *src*. A piece of wallpaper was left trailing onto the carpet.

That radiating gene could be used as a probe, a way to poke and prod DNA from other sources for a matching gene sequence. They knew *src* had come from somewhere outside of the viral genome. The probe gave them a way to search for it in other genomes, because only an actual *src* gene would match up with the *src* probe. Like a detective with a photograph of a missing person, they could hold the probe up to DNA from one species or another and search for the gene.

It took them four years to create the probe. The next step was to send out the search, using their radioactive *src* to find the gene outside of the virus. The first obvious genome to investigate was the chicken. After all, that was where the virus had been found in the first place. "To my utter astonishment, there it was," said Bishop. "And then we found it in duck DNA, and then we looked at more ancient birds, like ostrich, and it was there."

The discovery, made in 1976, was absolutely startling. Here was a virus that had somehow taken on a gene that caused cancer in chickens. And where had the gene come from? The chicken. What's more, the gene was present through a long trace of history. "Here it is preserved through all of evolution from the early metazoan on," said Bishop, who is still at the University of California–San Francisco forty years later. "It's a normal gene." So it turned out that *src* was a legitimate gene present in many species, which, if transposed into the genome of a retrovirus, became an oncogene.

The attempt to find the original source of the *src* gene had started as part of the quest to understand how *src* turned normal cells cancerous. But in tracing the gene back to its source, Bishop and Varmus inadvertently transformed the entire field of oncology. No one had imagined that cancer-causing genes came from the cells of the same species that later got the cancer.

The surprising discovery gave birth to two crucial notions that quickly became fundamental to the research of cancer and its treatment, the wagon wheels on which the future of cancer medicine would be pioneered. First, Bishop and Varmus proved without a doubt that normal genes can change into cancer-causing genes. How and why that change occurred remained a mystery, but the fact itself was undeniable: A normal gene that goes about its business in the DNA of a healthy animal can, without notice, turn into an oncogene, somehow launching a series of cellular events that lead to cancer.

But that wasn't all that the *src* probe research showed. More shocking was Bishop and Varmus's proof that these normal genes with oncogenic potential—proto-oncogenes, as they became known—were part of the animal that got the cancer. Until this moment, most scientists had assumed that the oncogene had come from the scene of the crime: the moment RSV had infected the chicken. But Bishop and Varmus traced the criminal back to its roots, to the time when it was still innocent. The *src* in RSV was an oncogene, but this same gene existed in the normal cells of healthy chickens. The oncogene had originated not in a virus but in a mammal cell. At some point in its history, RSV had infected a chicken and picked up the *src* gene from the chicken as it replicated inside its host. But when *src* was incorporated into the viral genome, it had mutated into a cancer-causing form of itself. Thus was born the understanding of the cellular origin of oncogenes. The phenomenon of a virus taking on a proto-oncogene from its host was dubbed "oncogene capture."

But was the normal *src* gene present in other mammals, such as mice and—the ultimate question—humans? As Bishop and Varmus widened their net to allow for more diversity with their probe, they found that normal, healthy, non-cancer-causing *src* was in the human genome. So a gene known to cause cancer in the animal from whence it had originally come was also present in human DNA. Might cancer in people be attributable to the same mechanism? Did humans have proto-oncogenes that could, somehow, turn oncogenic? If so, how would that change occur? In 1989, Bishop and Varmus would win the Nobel Prize for their discovery of "the cellular origin of retroviral oncogenes." They had proved the seemingly impossible: Cancer-causing genes in retroviruses were once normal genes in normal cells. This work transformed all of cancer research to follow. Eventually, the journey to understand oncogenes would lead to the genetic roots of cancer.

THE IDEA—RATHER, the fact—that *src* was a normal cellular gene that had, at some point in history, been scooped up by RSV and, through that adoption, been rendered oncogenic, brought a whole new level of importance to Erikson's search. Understanding the function of the gene was more important than ever. If cancer genes were altered forms of normal genes, then knowing what those genes normally did could prove crucial. It might allow scientists to identify other proto-oncogenes. It might provide clues about why the alteration occurred. Maybe, eventually, it would allow for cancer to be prevented or at least treated more effectively. Pinpointing the cellular mechanism associated with a proto-oncogene could mean pinpointing the cellular mechanism that was hijacked to induce cancer.

Bishop, too, began looking for the protein product of *src*. The question was so obviously central to understanding cancer: If the transformation of this single gene was enough to cause cancer, then knowing what protein that gene encoded was vital. Understanding where the oncogene had come from was cause to celebrate. But why did the change from proto-oncogene to oncogene occur? And how did the oncogene do its bidding? That the gene encoded a protein was clear; most genes did, according to the science of the time. But no one had a clue about what type of protein was involved.

Proteins are made of strings of amino acids, a family of twenty-two molecules inherent to all life, most of which occur naturally in our bodies. Proteins shape how we look, determine the speed of our metabolism, and conduct every minute process constantly churning inside our cells. Discovered in the early 1800s by a German scientist named Gerrit Mulder, proteins—a word derived from the Greek *protos*, meaning "first"—are considered the basic substances of all living organisms; they translate our genetic code into a physical, tangible result. Egg whites were the first protein Mulder identified, and the catalog grew exponentially from there.

Knowing the pedigree of the *src* gene did little to aid the search for the protein it encoded. There were scores of proteins to choose from and still few clues about the criminal's weapon of choice. In 1976, six years after Erikson had begun his search, he did not publish a single scientific paper, and he was beginning to wonder about the future of his lab. Without any publications to show for his work, grant money would grow harder to come by. "There were some very good students there," recalled Erikson. "If I couldn't acquire funds, they would have to be let go."

Finally, in 1977, Erikson and one of his students, Marc Collett, found the protein product of *src,* a discovery that generated excitement even among colleagues who'd teased Erikson for focusing his energy on cancer in chickens. But the satisfaction of the breakthrough was fleeting. After all, as nice as it was to be able to point to a protein and know that this was the one made by *src,* that information was actually fairly useless. What science really needed to know was the function of the protein. What did it actually do in the cell? Was it responsible for cell division? Did it help power the cell? If the cell is like a house, then identifying *src*'s product was like pointing to a single part of the house and declaring it important. Identifying its function meant knowing whether you were pointing to a fuse box, a radiator, or a water pipe. A year later, Collett and Erikson had figured it out. The product of *src* was an enzyme called a *protein kinase.**

Finding a kinase at the other end of *src* came as a surprise to Erikson and Collett and every other scientist studying viruses and cancer. Kinases are enzymes, a class of proteins that act as catalysts, facilitating processes in the cell that are essential to survival. In the years following Erikson and Collett's finding, more than 500 kinases would be found inside the human body. They would eventually be found in nearly every living species (paramecia, those single-celled organisms usually seen on our first glimpse through a microscope, hold the world record with about 2,600 kinases). But in 1977, only a few had been identified. And as far as anyone knew, none of them had anything to do with cancer.

Much of how kinases operate had been unraveled by Sir Philip Cohen at the University of Dundee in Scotland. There was only one known kinase when Cohen entered the field in the 1960s, and his lab had been largely responsible for illuminating the role kinases played in "all aspects of how living cells work," he said. The term "kinase" itself hinted at their importance: The root is the same as in "kinetic," from the Greek

---

* Bishop, who, together with lab member Art Levinson, had likewise been searching for the *src* gene product, also found the kinase. Their paper was published three months later. Credit for the discovery goes to Erikson and Collett.

kinesis, meaning "motion." What Cohen and others had found is that most cellular processes occur in a cascade of events. Insulin, for example, triggers one signal, which triggers another and then another. Charting these signaling pathways—signal transduction, the chain would come to be called—for a given process is a hallmark of the basic research behind understanding human diseases. Kinases, Cohen was finding, were almost always at the starting point of that cascade.

What's more, biologists had managed to figure out the exact mechanism that kinases use to set those pathways in motion, the flag the enzymes wave to start the race. Inside every living cell float molecules of ATP, or adenosine triphosphate, the container of energy inside our cells. The adenosine is a sugar-based chemical that serves as the backbone for three molecules of phosphate, hanging like beads on a string, with powerful bonds holding the beads together. ATP is, essentially, fuel.

To launch a signaling cascade, kinases pluck a phosphate from ATP—one bead is removed from the string—and stick it to a protein. The phosphate wakes up the protein, which immediately starts performing whatever job it's encoded to do. Once those duties are complete, another enzyme comes along and removes the phosphate, and the protein settles back into its more dormant state.

"Kinases deliver the right materials at the right moment to the right place in a cell at minimum energy cost to a cell," said Cohen, who has been knighted twice for his contributions to science. The process of delivering a phosphate to a protein, known as protein phosphorylation, is incredibly efficient, launching a chain of crucial events in one swift motion. The kinase, it turns out, is a model of sustainable energy.

It is also, as scientists would soon discover, the perfect tool for cancer to proliferate.

In the 1970s, Cohen's lab had become a magnet for kinase-inclined scientists. He'd been a lead researcher in parsing the role kinases played in converting glucose, the sugar into which food is converted in our cells, into glycogen, the stored form of that sugar. After work, these scientists, mostly young single men, would tromp through the cold Scottish evening to a now-long-gone pub on the Hawk Hill Road to hash over their research. They were often accompanied by Nick Lydon, a postdoc from the next lab down the hall.

Lydon had never been considered particularly brilliant. He was dyslexic, and so he never excelled in his classes at the Strathallan boarding school in Perth, Scotland. With its lack of prose to wade through, science proved much easier for him than literature or language. Because in those days the UK educational system required students to choose a specialty early on, Lydon ended up sticking with the hard sciences. He attended the University of Leeds, where he majored in biochemistry and zoology, and followed his undergraduate degree with a PhD in biochemistry. He arrived at the University of Dundee for postdoctoral training in 1978, just as Erikson was rattling the field with his report about the Src kinase.

Lydon was studying hormonal regulation, but the work lacked the bite of practical relevance. By contrast, the sounds of science emanating from the Cohen lab piqued his interest to no end. Lydon, his piercing blue eyes reflecting his active mind, began hovering around the lab benches of Cohen's postdocs—Brian Hemmings, Peter Parker, and Colin Picton, in particular—to soak up all he could about how kinases were involved in glucose metabolism, the breakdown of glycogen, and all other aspects of the signaling pathway that began with insulin. He sat in on seminars held by the lab and regularly joined Hemmings and the rest for pints.

The potential for this research to be useful in a tangible way excited Lydon. "I'd always been interested in application more than basic research," said Lydon. And there was something about kinases that he couldn't ignore, like the lamp from a distant lighthouse on some unknown shore. As he chatted with Cohen's postdocs about the field, he heard about the discovery that the cancer-causing gene *src* encoded a kinase. It was the first time this family of enzymes had been found in cancer. Lydon had to know more. Their involvement in the insulin pathway was intriguing enough. But for a scientist wanting to make sure his work had a practical application, studying a protein involved in cancer seemed a gold mine. Gradually, after watching from the sidelines, Lydon knew he wanted to turn his focus to kinases. Figuring out how he would do that was another matter entirely.

6

CONSUMMATE INSTIGATORS

Bishop and Varmus's discovery of the cellular origin of the *src* gene spurred a chase for other similarly derived oncogenes. If RSV achieved this odd feat, could other viruses do the same? As luck would have it, the technology to do an expansive, rapid search for oncogene candidates became available at just the right moment. In the early 1970s, a group of graduate students at Stanford University discovered several different enzymes that allowed them to join together DNA from different sources, recombining it into a single strand. Recombinant DNA, as it came to be called, enabled scientists to investigate individual genes in a way much easier than creating radioactive probes. The technique became so integral to so many aspects of science that it even seeped into mainstream knowledge, but under its more easily grasped name: cloning.

The invention of cloning in the early 1970s was nothing short of a revolution for the field of genetics. It allowed for individual DNA molecules to be propagated in a cell culture, and it allowed viral DNA to be duplicated outside of the natural host. Most relevant to this story, recombinant DNA technology enabled scientists to take apart viral genomes and test the cancer-causing potential of each individual gene. Viruses usually comprise just four or five genes, and with recombinant DNA, each of those few genes could be cloned individually, revealing which single gene was responsible for inducing transformation. It was like lifting up each cup to find out which one hid the magician's coin.* This line of inquiry—Were there other oncogenes like *src* out there?—quickly bore fruit. The list of such proto-oncogenes continued to expand: *myc, ras*, and, later, hundreds of other human genes. A few years later, in the early 1980s, some scientists made another conceptual leap: If a virus can turn a normal gene into an oncogene, could environmental factors do the same thing? "Why wouldn't cigarette smoke be able to do that? Why wouldn't sunlight be able to do that?" said Bishop, recalling the questions of the day. The notion that a proto-oncogene could be converted into an oncogene by something other than a virus was taking root. No one was sure what the exact trigger might be, or how to even search for the trigger. But slowly, oncogenes were pried loose from their historical ties with viruses. The research into oncogenes was taking on a life all its own.

In 1977, when Erikson and Collett pronounced that the product of the *src* gene was a kinase, these ideas remained almost entirely unimagined. Many scientists working in this vein assumed that the discoveries they were making were relevant to human cancer, but there was no concrete link. Making the leap from proto-oncogenes in chickens to proto-oncogenes in people meant crossing an enormous chasm.

The fact that Src was a kinase had come as a complete shock. Only a handful of kinases had been found in human cells, and they were assumed to be involved in only a few very specific processes in the cell. What on earth could they have to do with cancer? The discovery brought instant recognition to Erikson and Collett and launched a frenzy of new research about kinases. "It was quite a zoo for a while," recalled Erikson. The search was on to figure out whether any other cancer-causing viruses were also abusing the natural kinase mechanism.

* The technology behind recombinant DNA is credited to Herb Boyer, then at the University of California–San Francisco, and Stanley Cohen, then at Stanford University. In 1976, Boyer and venture capitalist Robert Swanson founded Genentech, the first biotech company, with the goal of using recombinant DNA to develop new drugs. The company's first product, human insulin, was cloned in 1978. The pharmaceutical company Eli Lilly licensed the drug, which was approved in 1982.

If scientists were at first baffled by the appearance of a kinase on the stage of cancer research, the logic of their involvement slowly became clear. As research continued to pull back the curtain, kinases were found to have roles in all manner of cellular processes: cell division, metabolism, insulin breakdown, and numerous other pathways and actions that are essential for living. As consummate instigators, setting off cascades of signals by the simple bestowal of a phosphate onto a protein, kinases suddenly seemed like the obvious place for cancer to begin. This kind of general mechanism, with the power to guide so many crucial functions, science would later realize, is perfectly suited for the development of cancer, a disease of unchecked cell growth. If cancer is like a gun, then the kinase is like the finger behind the trigger.

But so much mystery still remained. Normal cells had normal *src* genes and, therefore, normal Src kinases. When the Rous sarcoma virus picked up the *src* gene from a chicken and integrated it into its own genome, the gene became an oncogene, an abnormal version that caused cancer. Clearly, then, the protein encoded by the gene—the Src kinase—was also abnormal. An onco-protein, so to speak. But how? What was the difference between the Src kinase in healthy cells and the Src kinase in cancer cells?

JUST AS THE world of cancer treatment was about to enter its modern era, Brian Druker was entering medical school, where he would soon be forced to admit his interest in cancer, something he'd tried to deny since his first undergraduate years at University of California–San Diego. He'd always done well in school; academic excellence had always been strongly encouraged in the tight-knit household where he and his three siblings had grown up in St. Paul, Minnesota. But he'd resisted pigeonholing himself into medicine, and his applications to med school had lacked the hallmarks of the students who'd shaped their every decision around becoming a doctor, their résumés littered with hospital volunteer hours and early anatomy courses. Rather than marching steadily into the medical profession, Druker had stumbled forward one step at a time, charting his course according to what was appealing to him at the time. He'd delved into laboratory research out of curiosity, and when the time came to apply to medical school, he'd amassed far more hours in the lab than at a hospital, a track record he assumed would be a strike against him. So it came as a surprise when his interviewer at University of California—San Diego Medical School, a man named Russell Doolittle, who was one of the most highly regarded scientists on the campus and known to be impossible to please, told Druker at their first meeting, "You're exactly the kind of person I like to see in our medical school." At the time, Druker was just relieved for some recognition, having bumbled his way through his previous interviews. Only later did he realize the future that he'd secured for himself by following his interest in the lab bench. "It was the path I had chosen for medical school that he wanted to assist," Druker said.

Still, for the moment he was keeping quiet about his interest in cancer. "You have to realize, cancer was a devastating disease," said Druker. "Everybody died." The disease was a grim, dark domain where only the most morbid physician dared tread, and Druker was unwilling to admit, even to himself, that he was fascinated by it. "Everybody was afraid of it, and people in oncology [were] weird because this disease was so hopeless," he recalled. "Why would you go take care of patients with no hope? You were crazy if you were going to do that."

7

WHERE THE KINASE HANGS THE KEYS

Just a year after the Src kinase was discovered came the next significant moment in kinase research. While Erikson was working out the *src* product, Tony Hunter, a British scientist doing his postdoctoral training at the Salk Institute in La Jolla, California, was examining a virus known as polyomavirus, which causes multiple types of tumors in rodents. The DNA sequence of the virus was just being made available. Just as with the Rous sarcoma virus, polyomavirus included some genes that could render cells cancerous and others that allowed the virus to successfully infect its host. By the late 1970s, Hunter knew that those responsibilities were equally distributed among the polyomavirus genome, with about half of the genes related to replication and the other half related to cancer transformation. Hunter had been trying to understand how the cancer-related genes induced cancer. It was the same question Erikson had been asking: What was the protein product of the gene, and how did that protein spur tumor growth?

By 1978, Hunter had figured out that the polyomavirus made three proteins that were responsible for causing rodent cancer. Elsewhere, researchers studying the same virus showed that one of these proteins, the middle one, known as "middle T," was behind the conversion of skin cells into tumor cells. Now the question was: What kind of protein was it? It was the same question that had led Erikson on an eight-year trek down a rabbit hole.

When Erikson and Collett's Src kinase report came out, Hunter immediately wondered whether his polyoma protein was the same variety. "Maybe it was a universal mechanism of transformation," Hunter recalled thinking at the time. And that was exactly what he and two other research groups found: The crucial protein was a kinase. "That was obviously incredibly exciting," says Hunter. "It pretty much proved that kinase activity was going to be important, although we didn't know how."

But his restless mind wasn't done asking questions about this protein. As the role of kinases in cancer was becoming more apparent, Hunter wanted to know how, exactly, they accomplished this devilish feat.

Over several years, Phil Cohen, at Dundee, and others had parsed the process of phosphorylation, the crucial placement of a phosphate by a kinase onto another protein in the cell to catalyze a stream of microscopic events. Scientists had figured out that when the kinase delivers a phosphate to a protein, the phosphate can't land just anywhere. Just like having a spot in a house where the keys are kept, the kinase needs a specific target location on which to deposit the phosphate it removes from ATP, the triplet of phosphates that serves as the main storehouse of energy within the cell like a battery.

That target location, the place where the kinase always hangs the keys, was a single amino acid, one of the many strung together to form a particular protein. Only certain amino acids can serve this function. During the time that Hunter was investigating polyomavirus, two amino acids were known to be the targets: serine and threonine. Every known kinase placed phosphates onto either serine or threonine.

Once Hunter had confirmed that the protein responsible for the cancer induced by the polyomavirus was a kinase, he wanted to connect the dots further, to describe exactly how and why that kinase caused cancer. As a first step, he planned to check which amino acid the kinase phosphorylated, serine or threonine.

To find out which amino acid was the target of any kinase, researchers immersed a protein in a solution that separated out the amino acids at a pH of 1.9. When the amino acids were separated, it was possible to see which one carried the phosphate. If the solution got old, though, and a lazy researcher didn't refresh it, the pH changed, and different amino acids would emerge.

Hunter was feeling particularly lazy one day in 1979, and he used old buffer solution as he continued his search for the amino acid phosphorylated by the middle T kinase. One of the amino acids that emerged as a result was tyrosine. Tyrosine wasn't a very popular amino acid in scientific research, and Hunter had hardly been on the lookout for it, especially considering that every other kinase known at that time brought phosphate only to either serine or threonine. But when he took a closer look, he saw that the tyrosine was phosphorylated.

Hunter was taken aback. It was like hearing a foreign language for the first time and realizing that there are other ways to communicate. Tyrosine kinase is exceptionally rare in the human body. Of the more than 500 kinases known today, only about ninety attach phosphates to tyrosine. And those ninety kinases activate less than 1 percent of the proteins in the human body. In other words, fewer than ten of every 1,000 proteins in the body have tyrosine as their receiving dock. The other 900-plus proteins use serine, threonine, or a combination of both. Hunter didn't know just how rare tyrosine phosphorylation was at the time, but he knew he'd stumbled onto an oddity. Of the several kinases known at that point, all stuck their phosphates onto serine or threonine. Why would this one target tyrosine? And was that fact important?

Later, Hunter and other researchers realized that the middle T protein he'd been studying was not actually a purified single protein, but rather was contaminated with other proteins, a common occurrence in lab research at the time. It wasn't middle T that was the kinase, it was the contaminant. But that realization didn't alter the shock of finding tyrosine on the receiving end of a phosphate. Hunter's discovery had turned everyone's attention to this once-ignored amino acid called tyrosine, and that was what mattered. A new concept had been born: Tyrosine kinases were major determinants in driving the growth of cells, including malignant ones. This understanding would prove to be a major step forward for cancer researchers. In the hunt for cancer's underlying mechanism, finding tyrosine was like a tracker finding an animal's footprint or a broken branch. They knew they had caught the right trail.

BACK IN SCOTLAND, Lydon continued his postdoctoral work at Dundee and continued to absorb all he could about kinases from the lab next door until 1982. By that point, having spent most of his life in Scotland and northern England, he was ready for warmer, if not greener, pastures, and he accepted a job at Schering-Plough, the Paris-based pharmaceutical company. His boss was a medical doctor from Switzerland named Alex Matter.

As Lydon was finishing up at Dundee, Druker was in his first year of medical school and taking a course on the history of cancer therapy. He slowly began musing about what it would take to improve cancer treatment, wondering whether it might be possible to create medicines that affected cancerous cells differently from how they affected normal cells.

Like pieces on a chessboard assembling for checkmate, discoveries and researchers were slowly coming together. The mutant chromosome, the translocation, the tyrosine kinase. Nick Lydon, Alex Matter, Brian Druker. The next advancing move came from a colony of mice in Bethesda, Maryland.

8

A CHEMICAL AMPUTATION

In 1965, Herb Abelson was in his mid-twenties and had just graduated from medical school, and he had a serious problem. He was a prime candidate to be drafted to serve as a general medical officer for the Vietnam War. But Abelson was vehemently against the war and had no desire to be shipped off to the jungles of Southeast Asia. Yet he also had no desire to flee to Canada. The solution was to join either the Coast Guard or the National Institutes of Health.

During the war, the National Institutes of Health (NIH) became a refuge for smart medical doctors like Abelson who did not want to serve overseas. Employees at the NIH were part of the Public Health Service, which was considered a military position. A doctor working there could safely avoid being sent to Vietnam. With approximately 15,000 med school graduates and only 100 NIH job openings, it was the cream of the crop that ended up there. This hotbed of talent ended up making enormous strides in the treatment and understanding of cancer, as well as many other diseases. Considering the circumstances leading to his NIH arrival, the contribution Herb Abelson was about to make to the understanding of how CML develops and the leap in cancer treatment that ultimately resulted from that explanation could easily never have happened.

Fortunately, his arrival at the NIH also coincided with an era of solid support for young scientists with new ideas. In today's risk-averse climate, the average age of a first-time National Cancer Institute grant recipient is in the early forties, with younger applicants routinely denied funding to investigate new ideas. In 2010, less than 4 percent of R01 grants—the oldest and most common NIH award, for up to $250,000 per year up to five years—went to scientists under age 36. The average age of R01 recipients has continued to increase during the past thirty years, even as the average age of newly minted PhDs has remained constant. But conditions and caution were quite different in the 1960s and '70s. Abelson, like Nowell and Hungerford, was young at a time when youth was considered an advantage in making medical breakthroughs.

Abelson was assigned to work under a biologist named Jack Dalton, who had a hands-off style that left his charges free to pursue whatever struck their curiosity. Like Rous fifty years earlier, Abelson was particularly curious about viruses that caused cancer.

The world already knew a lot about viruses by the time Abelson came to the NIH. Most crucial here was the fact that although viruses, which can't reproduce on their own outside of a host, don't qualify as living things, they do have genetic material. Once a virus finds its way inside a living being, be it plant, animal, or human, that genetic material is replicated over and over. Sometimes, such as with HIV, the virus that causes AIDS, that replication, left untreated, eventually kills the host. Other times, the body eradicates the virus on its own, as is usually the case with the influenza virus. Sometimes, the host grows accustomed to a virus, its presence proving undisruptive to the host's normal goings-on.

But many other crucial facts had not yet come to light when Abelson joined the NIH. Temin had not created his focus assay or found the hidden mechanism that allowed RNA viruses to replicate. Erikson had not yet identified the protein made by the killer gene in RSV, and Hunter had not uncovered tyrosine's involvement in cancer. No one had a clue about the cellular origin of oncogenes. The chromosome banding techniques that would allow Rowley her seminal observation were nonexistent. These three threads of research—the Philadelphia chromosome, the role of kinases in cancer, and the source of the cancer-inducing gene in cancer-causing viruses—remained utterly distinct. And the research Abelson was about to begin was another completely unrelated strand of science. He was interested in pulling apart the process by which viruses caused cancer, charting their course inside a cell. That investigation wasn't necessarily about genes. For Abelson, it wasn't necessarily about anything more than asking a question and seeing where his attempt to answer it would lead.

Elsewhere at the NIH, other draft avoiders were experimenting with toxic chemotherapy regimens, desperate for some step forward in the treatment of cancer. Cancer treatment had gained significant ground since the introduction of corticosteroids (synthetically manufactured versions of hormones produced by our adrenal glands) and antifolates (drugs that thwart folic acid, a chemical essential to the production of DNA) in the 1940s. These chemicals extended the survival time for many cancer patients, though always incrementally. Any attempt to improve treatment by combining drugs was trial by error, trying different drugs given in varying regimens and waiting to see what happened. At the NIH, determined doctors injected patient after patient with one concoction after another, hopeful that each new mix would be better at killing the cancer without killing the patient. Behind thin curtains, patients moaned in their hospital beds, either from the pain of the cancer expanding in their bodies or the toxic side effects of the drugs. Though strides were made during those years that would help steer decades of cancer care to come, frustration was much more common than success.

Far from those hallowed halls, Abelson was busy at the lab bench, where he homed in on the Moloney virus, which caused leukemia in mice. Moloney virus is what's known as a simple RNA virus. Simple RNA viruses, which often cause cancer, come in two varieties. One type contains all the genes required for replication. These viruses often cause tumors in animals, and are found most commonly in domesticated animals such as cats, food animals such as chickens, and laboratory animals. Almost always, the virus is found because of the tumor it causes; pet owners, farmers, and scientists see the lesions, not the infection. The viruses do exist in the wild but are encountered less frequently because the close contact required to spot them is much rarer. These tumors usually have a long latency period; they are very slow to induce disease.

The second type of simple RNA virus is less common but carries a more immediate danger to its host. These are the viruses that, when they replicate, capture genes from the host. By the time the virus infects a new host, though, the captured gene has become an oncogene. RSV was an example of this simple RNA virus. The virus had captured *src* from a host during replication, and once inside the viral genome, the gene had become an oncogene that caused cancer in chickens. This transport, a hostage from one ship who turns criminal by the time of the next piracy, was the process that Bishop and Varmus would spot with their radioactive DNA flashlight, the *src* probe that had enabled them to find the normal, proto-oncogene version of *src* in the genome of healthy chickens, the very animal later infected by a virus containing a cancer-causing version of *src*.

When Herb Abelson was deep into his lab work at the NIH in the 1960s, he knew nothing about this. The concept of oncogene capture was years away. In fact, the very notion of oncogenes—healthy genes with the capacity to turn malignant—was a completely foreign concept to lab researchers like Abelson. He knew about the common simple RNA viruses; he was using one in his experiments. He knew nothing about the rarer kind that could lift a gene from its host and thereby trigger a deadly disease. But he was about to witness one taking over his colony of mice.

EVERY VIRUS HAS a target cell type, a certain organ that it most wants to infect. Abelson knew that the Moloney virus hit the thymus gland first. From there, a type of leukemia called lymphocytic leukemia—a cancer of the T-cell portion of the immune system—spread to the rest of the body.

Abelson wanted to see if he could get the virus to target a different part of the mouse's body—if he could, as the science jargon goes, "increase the host range" of the virus. If the road was blocked, would the driver find an alternative route or just call it a day? The reason for the experiment was simple: He was curious. He had an inkling that testing whether he could force the virus into a new path of attack might show him something about how cancer happens, though he wasn't sure what.

He injected 163 newborn mice with prednisolone, a powerful steroid that shrinks the thymus and other parts of the immune system. He was chemically amputating the thymus from the mice. Then, when the animals were between one and eight weeks old, he injected them with Moloney virus. Without the thymus present, what kind of cells would the virus gravitate toward? Would it replicate in another part of the body, or would it fail to infect, leaving the mice cancer free?

In the days that followed, Abelson visited his mouse cages looking for signs of cancer, his heavy brow still as he took each animal in his hands to peer through its fur. Leukemia developed in more than 100 animals. Twelve of the mice developed lymphosarcomas, hard tumors in their lymph nodes, where lymph, part of the immune system, gathers. "That was pretty unprecedented," said Abelson. "These large tumors had never been seen." The cancer was affecting the B cells of the immune system instead of the T cells, the original target. What's more, unlike the slow-growing cancer caused by the Moloney virus, this malignancy developed rapidly, with a startlingly short latency period. All of the animals were dead within weeks.

B cells and T cells are lymphocytes, one of the varieties of white blood cells that make up the immune system. Both B and T cells have receptors on their surface that recognize foreign substances, called antigens, with each receptor tuned to a different antigen. But though their functions overlap, B and T cells are distinctly different. The majority of T cells are either helpers or killers, the helpers triggering the killers into action to annihilate invaders like viruses, some bacteria, and some cancer cells. Helper T cells can also activate B cells, which produce bacteria-fighting antibodies and new immune cells that remember new invaders. B and T cells also reside in different parts of the body. B cells grow entirely in the bone marrow, whereas T cells, though sprouted in the marrow, mature in the thymus. Their differences in function and location are such that a virus that causes cancer in T cells does not necessarily have the same ability in B cells. So finding B-cell cancer in mice injected with a virus known to cause T-cell cancer was very odd.

Abelson wanted to know more about the anomalous hard nodules that had grown in the lymph nodes of some of the mice in his colony.

Was the virus inside the mice still Moloney or was it something different? Was the virus able to cause a different type of cancer when its usual pathway was thwarted, or had the virus itself been altered when it infected the chemically amputated mice, changed irrevocably into a deadly new virus that killed mice far more rapidly than its predecessor? To find out, he would inject that extracted virus into new mice—thymuses intact—and see what kind of cancer they got.

Abelson extracted tumors from a few of the lymphosarcoma mice and passed them through filters so fine that only the virus would pass through. Abelson infected a new, healthy colony of mice with the purified sample. Then he waited to see what kind of cancer would develop. If they got slow-growing thymus cancer, Abelson could be fairly certain that the Moloney virus had come through the experiment unscathed, and that, simply, in the absence of a thymus to land on, the virus had entered another cell type; it had taken the alternative route. But if the healthy mice still got tumors in their B cells—the type of immune cell in which cancer had arisen in the chemically amputated mice— then the virus couldn't possibly be the same one he'd started with.

The answer came quickly: All of the mice developed B-cell tumors. Again, they were dead within weeks. Clearly, the agent he'd injected into the second colony wasn't the Moloney virus. It was something different, some altered version. Abelson had found a new virus.

Eventually, the Abelson virus, as it came to be called, would play a central role in the story of the Philadelphia chromosome and CML. The virus was a valuable research tool not so much because of the cancer that it caused but rather because it gave scientists a controlled way to study how cancer develops. There were three key features that made the virus the perfect model for exploring cancer transformation. First, its pedigree was known, so there were no questions about where the virus had come from that might cast doubt on promising research findings. Second, the virus caused cancer that progressed rapidly, allowing experiments to be completed in weeks or months rather than years. Third, the cancer it caused was one of the blood, easier to study than solid tumors of internal organs because they can be extracted with syringes rather than with surgery.

In 1969, when Abelson presented his work at the annual meeting of the American Association of Cancer Research (AACR), the significance of the virus was wholly unsuspected. In fact, the response belittled his findings. Some more seasoned researchers criticized Abelson for playing around in the lab when he should have been dutifully studying the scientific literature and following a more well-worn path. "I was berated by the president of AACR, and disparaged about it," Abelson recalled.

But much like David Hungerford with his predilection for recording his observations simply for the sake of recording them, Abelson wasn't worried about where his work was leading to, and neither were his supervisors or those providing the money. It was enough that the science was solid, that it answered some questions and raised new ones.

And it turned out that Abelson wasn't really interested in pursuing those questions any further. Rather, he found himself gravitating toward treating pediatric cancer, and so, in the early 1970s, he left the NIH to complete his medical training at Boston Children's Hospital.

He still yearned for the excitement of laboratory research, though, and didn't want to cut himself off completely from that kind of work. So he joined a lab at MIT, across the river in Cambridge, where he worked during his residency. He stored away his samples of the virus that now bore his name and moved on to other pursuits.

9

STRIPPING AWAY THE FUR AND THE FAT

One floor below Abelson's new lab bench, David Baltimore was feeling impatient. Fresh off his and Howard Temin's groundbreaking discovery of reverse transcriptase—the enzyme that enabled RNA-based viruses to convert their genetic information into DNA that could then be replicated inside a host—he was eager to put the finding to good use.

Locating the mechanism that enabled RNA viruses to replicate portended a new era for understanding cancer because it created the ability to investigate those viruses at the molecular level. "I was convinced that mouse viruses were going to make the difference, that the ability to manipulate the genetics of the mouse gives you a research range that you can't get any other way," said Baltimore. "But having had no experience with such viruses, I didn't have a system to work with, and I didn't know what system would be best." Each step forward in this uncharted terrain meant inventing new steps, new ways to grab hold of the landscape, new methods for tilling the earth. Cloning held so much possibility for studying how cancer occurred, but researchers had to find ways to exploit the technology first.

Whereas the Colorado-based lab of Ray Erikson—the man who, with Marc Collett, had discovered the Src kinase encoded by the cancer-causing *src* gene in the Rous sarcoma virus—was infused with the slow pace of his country upbringing, leaving him contented with taking seven or eight years to answer a question, Baltimore, born and raised in New York City, valued quick results. "Life is too short to sit around for years and years to get an answer to anything," said Baltimore. "Simplicity and speed are things you don't hear a lot about, but to me, they are central elements to research." Now determined to exploit reverse transcriptase as a cancer research tool, Baltimore—whose success had already brought him renown, with the funding and resources to prove it—stocked his lab with postdoctoral talent that would help him gain ground quickly.

When Naomi Rosenberg joined the Baltimore lab in 1973, she wasn't particularly interested in reverse transcriptase. The first person in her rural Vermont family to attend college, Rosenberg's early childhood interest in science had withered during boring high school classes. Later, having come to terms with the slim career options afforded by her chosen college major of Latin, she decided to give biology another try. She was hooked. Veering far from the paths of her cabinetmaker father, pottery-making mother, and poetry-writing brother, Rosenberg found her calling in the microscopic world of viruses. "The idea that something so small could have such devastating consequences—I just find it fascinating," Rosenberg said.

Baltimore didn't have a virus team at his lab, but he did have something else that greatly appealed to Rosenberg. "One of David's amazing strengths was the freedom that he gave everyone who came to his lab to work on whatever they wanted as long as it fit under a very huge, general umbrella," she said. His instruction to her upon her arrival was to figure out what she wanted to do. The lack of guidance was daunting yet irresistible to Rosenberg, who was unaccustomed to such openness in an academic lab. And her timing was perfect. A renovation of the MIT Cancer Research Building, supposed to have been completed by Rosenberg's arrival, was taking longer than expected. Temporarily left without a lab bench, Rosenberg had ample time to search the scientific literature and formulate a plan.

And Rosenberg, despite her lukewarm attitude about reverse transcriptase, had something else for which Baltimore had been searching. She possessed a unique talent in the world of 1970s virology: the ability to grow and manipulate different kinds of cells in culture. In the 1960s and early 1970s, simple RNA viruses were usually studied in cells that came from sources other than the tumors they caused. Examining the ins and outs of how a virus infected a random bunch of cells was definitely interesting. So much was still unknown about viruses, let alone how, exactly, they cause cancer, that any incremental finding mattered. But as long as the cells were not from actual tumors that had grown inside actual animals infected with the virus, the insights remained abstract, an intellectual exercise that, though a valid contribution to science, lacked a gripping connection to cancer in people or animals. Temin's focus assay, the breakthrough technique from 1958, had made the unseen visible by highlighting abnormal cells against a background of normal cells. But the cells were already cancerous. Rosenberg wanted to study the actual cells from animals infected with a cancer-causing RNA viruses in the process of cancer transformation before, during, and after the time they were infected. She wanted to create a system that would allow her to be the voyeur, watching cells become infected and turn cancerous. "If you really wanted to understand how the virus caused the cancer in the animal, you needed to understand how the virus interacted with that kind of cell," Rosenberg said. She had to find a way to get cells from a living animal to continue growing outside of their warm-blooded home. And Baltimore knew that if he was to make any progress in manipulating mouse genetics as a way to understand how cancer-causing RNA viruses caused cancer, he had to have a system for studying mouse cells undergoing that transformation outside of the mouse.

Rosenberg needed a virus that would turn cells removed from an animal into cancer cells in tissue culture (that is, in a petri dish, with a neutral medium in which to grow). With all the fur and fat stripped away, Rosenberg might witness the virus committing its crime. The experiment would still be just a model of how tumors developed, but it would be a much more accurate representation. Armed with her green thumb for growing new cell lines, Rosenberg went searching for a good virus candidate. Baltimore, eager to use his discovery of reverse transcriptase to investigate how viruses do their dirty work, encouraged her and joined in the search.

With time on her side, Rosenberg spent hours pouring over the scientific literature. Finally, she stumbled on Abelson's 1969 paper describing the mouse virus model he'd created. It was exactly what she was looking for: a virus of known origin and a rapidly developing cancer of the blood in mice. She wanted to use the virus to create a system in which to observe cancer transformation. When Rosenberg brought the model to Baltimore's attention, a look of recognition crept across his face. Why was that name so familiar?

It took them about two minutes to realize why: Baltimore had already caught wind of Abelson's work because Abelson had joined a lab right upstairs from Baltimore's when he'd left the NIH. When Rosenberg approached Abelson about using his virus, he was happy to oblige, supplying her with a glass vial of ground-up, filtered tumor extract from a mouse that had been injected with the mysterious virus.

Rosenberg teamed up with a scientist named Chuck Scher, who'd already begun working with the virus. The goal of their work was to figure out which cells the virus infected in mice, find a way to grow those cells in tissue culture, and then watch those cells turn cancerous after exposure to the virus. By doing so, the scientists could, they hoped, chronicle exactly how healthy cells turned malignant.

Day after day, Rosenberg and Scher would test out different cells taken from uninfected mice. Would red blood cells infected with the virus continue reproducing in the semisolid medium of the petri dish? What about cells from their lymph nodes? The search involved more than simply sucking out cells with a syringe and squirting them onto a bit of inert agar. Rosenberg had to find not only the right cells to use from the mouse, but also the right process for moving them from the mouse into the culture, and for infecting them *in vitro*—that is, outside of the living organism—without killing them.

After several grueling months of this crash course in mouse research—Rosenberg had scarcely touched a lab animal before—and no cells cooperating with their wishes, Rosenberg and Scher tried a microscopic extraction of liver cells from mouse fetuses. After surgically removing the fetus from its mother, they dissected the embryo and transported a sample of cells from its liver onto the culture dish.

It worked. The cells became infected and continued to divide, and the virus continued to replicate. Like an impressionable child, these embryonic cells proved susceptible to the virus outside of the protective shield of the womb. And as they began to replicate, they turned cancerous. Bone marrow cells from mature mice, they later found, did likewise. Rosenberg had created a transformation system, as it was called, for the Abelson virus.

Now, Rosenberg and Baltimore had a way to observe normal mouse cells turning malignant. She would soon be joined by other postdocs in the Baltimore lab eager to parse the mechanism behind cancer transformation. What protein was responsible for catapulting cancer into action? What was the interplay between the viral genome and the mouse genome? These were all valid questions, and the Abelson virus had provided the perfect model for answering them, a replica of what went on inside cells. Rosenberg and her colleagues at the Baltimore lab weren't driven to understand the Abelson virus per se; after all, that virus caused an anomalous cancer in mice and wasn't relevant to human cancer—or so they thought. Simply, the virus gave them a way to watch the process. They thought that whatever process they observed in the cancer transformation system would apply to other RNA viruses that caused cancer, maybe all of them.

THAT WAS IN 1975, two years after Janet Rowley had shaken the foundations of genetics with her discovery of chromosomes swapping pieces. In the meantime, Erikson was making his way toward identifying the protein coded for by *src,* the key gene behind the Rous sarcoma virus. Bishop and Varmus were about to inform the world that the gene on that virus, known to cause cancer in chickens, came from the chicken cell itself. The inner workings of the Philadelphia chromosome, the search for the protein product encoded by *src*, and the search for origin of that *src* had now been joined by yet another unrelated pursuit. Or so it seemed at the time.

10

A FUNNY NEW PROTEIN

Rosenberg's transformation system was considered a major breakthrough at the Baltimore lab. Everyone there wondered what it might reveal about how cancer occurred. The possibility of uncovering the cellular mechanisms tapped by the virus to set a malignancy in motion was incredibly exciting. With Rosenberg having succeeded in creating a way to study the Abelson virus outside of a mouse, Owen Witte, another Baltimore lab postdoc, now began searching for the protein responsible for the transformation from healthy cell to cancer cell. The question guiding his search was the same one that Ray Erikson and Tony Hunter had asked: What protein was inducing cancer in the mice? But rather than starting with oncogenes contained in viruses, Witte had turned his attention to antibodies, his area of expertise.

Antibodies are proteins made by the immune system that can recognize foreign invaders and trigger an attack against them. The creation of experimental antibodies for lab research had been ongoing for some time. Antibodies automatically gravitate toward their designated antigens. So if a scientist wants to know whether a particular substance is present in a cell, an antibody can be used to track it down, like a metal detector scanning for gold lost underneath the sand.

The trick was making the antibody. Scientists knew the body naturally produced antibodies to foreign invaders. If an unrecognized protein is injected into an animal, the animal will likely produce an antibody to that protein. If the animal makes the antibody—it doesn't always happen—then the antibody can be extracted, manufactured on a commercial scale, and used to search for that same protein elsewhere.

Witte had become an expert at preparing antibodies while he was earning his PhD at Stanford University, studying other viruses that cause leukemia. After that, he returned to his New England roots as a medical intern at one of Harvard University's hospitals. His mother was dying of cancer, however, and Witte found he couldn't stomach such heavy exposure to illness. "It was not an easy time to be taking care of other sick people," he recalled. As it turned out, Witte preferred the bench to the bedside. "If you're taking care of sick people dying of horrible diseases, you don't really look forward to the next day as much as you do in the lab, where, frankly, every day is a potential for discovery." He left his clinical training and went to David Baltimore's lab in 1976, working across the bench from Steve Goff, who had also joined the lab as a postdoc. Considering his deep knowledge of protein chemistry, Witte's arrival could not have been timed any better. "It was a good time to do something I liked rather than something I was supposed to do," Witte said.

Witte began working with the Abelson virus after Rosenberg succeeded in creating the transformation system, which exposed the virus and the mouse cells. Witte thought he could use this system and his antibody techniques to find the culprit protein encoded by the oncogene in the virus.

His first step was to make an antibody. This antibody would help him identify proteins from the cells one by one so that he could take a closer look. The obvious problem here was that Witte didn't know what protein he was looking for—after all, that was the question driving his research—so how could he make an antibody against it?

He decided to use an antibody targeted against a protein called Gag, the product of the *gag* gene. Witte knew that *gag* was present in the Moloney virus, the one that Herb Abelson had initially injected into the chemically amputated mice. Since the Abelson virus was derived from the Moloney virus, the Abelson virus would most likely have the *gag* gene too. Witte also knew that viruses, including Abelson, have very few genes, just four or five, and therefore very few proteins.

So all he had to do was work through them one at a time to find the protein responsible for triggering cancer. He had a one-in-four or -five chance of choosing the key protein when he began with Gag. He could have started with another protein. The choice of Gag was a matter of convenience; it was the only one for which he had an antibody—as good a place to start as any.

Witte took an Abelson-infected cell—rendered that way inside Rosenberg's transformation system—and extracted all of its proteins. Then he added a tiny bit of Gag antibody. Witte knew that the antibody would immediately bind to its target. If the Gag protein was present in those cells, the Gag antibody would latch onto it. If Witte's antibody found its target protein, then he knew the approach was working, and he could begin homing in on which protein was behind the transformation of cells from normal to cancerous.

The experiment worked. The Gag antibody bound to a protein inside the mouse cells that had been infected with the cancer-inducing Abelson virus. Witte then coaxed the antibody and its target protein away from the other proteins in the mix by spinning the entire extract through a centrifuge. With the mixture split up into its various components, like shaken salad dressing settling into stripes of oil, vinegar, and spices, Witte could rinse away everything but the Gag antibody-and-protein pairing. He got rid of the antibody using detergent, leaving just the antigen, the Gag protein.

At the end of this process, Witte knew he'd obtained a purified Gag protein. Now he wanted to look more deeply inside the protein, to investigate all its molecules for any sign of a cancer-inducing tendency. Witte put the protein into a gel through which an electric current was run. The technique, called gel electrophoresis, caused the protein to separate into its component molecules. The result was surprising; the protein that Witte had extracted was not what he'd expected to find. It was Gag, but something else was going on. "It was a funny new protein," said Goff, who by then had joined the Baltimore lab as a postdoc. Funny because it was really big. The weight of molecules is typically represented in daltons, with the average protein weighing around 55 kilodaltons. This one weighed about 120 kilodaltons.

Witte and Rosenberg worked as a team to investigate the protein further. They had already made mutant strains of the Abelson virus that replicated inside cells but did not render those cells cancerous. When they injected the Gag antibody into those cells, they could see that the protein it attacked was smaller than the protein Witte had found in cells infected with the original strain of Abelson. The smaller size was much more in keeping with how they expected a protein to look.

Witte knew that the difference in size was significant. The size of a protein is determined by the number of amino acid sequences within. The mouse cells infected with the cancerous strain of Abelson had larger Gag proteins than those infected with the noncancerous strain. The amino acid sequences that accounted for the difference had to be responsible for triggering cancer transformation. What's more, the number of amino acids required to make up that difference in size meant that there had to be another protein attached to Gag. The "funny new protein" was really two proteins in one. Witte and Rosenberg dubbed the new protein—the extra bit attached to Gag—Abl, pronounced like "able," short for the virus to which it was connected. The combination protein that Witte had first extracted from the Abelson-infected mice became known as Gag/Abl.

It was an incredible coincidence that the Abl protein, the one responsible for triggering cancer in the Abelson-infected mice, happened to be stuck to Gag. It was only because he'd used the Gag antibody to fish out the Gag protein that Witte came across Abl. If it had not been fused to Gag, he would have missed it. Now, just like Erikson and Hunter when they found the proteins they'd sought, Witte wanted to identify the type of protein he'd found.

By then, Ray Erikson had discovered that the protein encoded by the *src* gene, the one that the Rous sarcoma virus used to cause cancer, encoded a kinase. Well aware of the pronouncement, Witte and everyone else in the Baltimore lab wondered if the excavated protein would prove to be the same kind of enzyme. With the reports from Bishop and Varmus about the cellular origin of oncogenes, and then from Erikson, the kinase was starting to seem like an almost obvious mechanism for inducing cancer in a host.

Over the years, scientists had found a way to use ATP, the basic unit of fuel inside cells, to determine the function of a protein. Witte knew how kinases worked: They took a phosphate from ATP and bound it to a protein. The fleeting presence of a phosphate triggered the protein into action, launching a cascade of signals within the cell that resulted in all manner of events: metabolism, division, blood cell production, and on and on.

Witte incubated the Gag/Abl protein with ATP in a test tube. If it was a kinase, then it would start grabbing phosphates from ATP and sticking them onto other substances. Sure enough, this was exactly what happened. Gag/Abl was a kinase.

Baltimore, still Witte's supervisor, knew the next logical step was to check whether the Abl kinase attached phosphates to serine or threonine, the two most common amino acid binding sites. Identifying this landing platform was part of the routine for kinase research. But when Witte ran the standard experiment to confirm which amino acid was in play, he quickly realized it was neither.

At almost this same moment, Tony Hunter made his discovery about the tyrosine kinase. The kinase involved in cancer induced by the polyomavirus he'd been studying phosphorylated not serine or threonine, but tyrosine, a little-known amino acid present in far smaller amounts than the other two. The splash made by Hunter's finding should have turned Baltimore and Witte's attention to that amino acid right away. Ray Erikson was quick to pick up Hunter's lead. He checked Src to see if it might phosphorylate tyrosine. His report that Src was a tyrosine kinase was published in 1980.

But the clue eluded Baltimore for the strangest of reasons. Other concurrent work in his lab had uncovered a link between poliovirus and tyrosine. Witte's kinase work came so close on the heels of that link that Baltimore assumed tyrosine could not be important in both research arenas. What were the odds that two ongoing efforts in his lab could both involve this obscure amino acid? "I put the possibility that this was tyrosine at a very low level, just because of the adage that lightning doesn't strike twice," said Baltimore. Finally, when tests for every other possible amino acid proved negative, they turned to tyrosine. The result was immediately clear: Abl—the protein encoded by the oncogene present in the Abelson virus—was a tyrosine kinase. With the discoveries that Src and now Abl were tyrosine kinases, the world of cancer research began to light upon the notion that tyrosine kinases—that is, kinases that bind phosphates to tyrosine as a way to power up a protein—might be fundamental to malignancies.

Still, although there were whiffs of some greater significance in the air, the research was not yet connected to cancer in people. "Is there any human disease like this?" Goff asked himself. "No idea. Not a clue." Witte was similarly clueless. "Whether it would have anything to do with human cancer, whether it would be a drug target, that was beyond the state of our knowledge at the time," he said.

11

THE FIRST SIGN OF A HUMAN CANCER GENE

On the heels of Witte's breakthrough, research on the Abelson virus continued. The discovery that Gag/Abl was a tyrosine kinase upped the ante dramatically. The Abelson virus had arisen spontaneously in a laboratory and caused a fast and fatal leukemia in mice. Now, the protein responsible for that cancer, Gag/Abl, was revealed to be a tyrosine kinase.

Two distinct strands of research—*src* and the Abelson virus—had now come together. Erikson's pursuit of the Src protein product had uncovered the role of kinases in cancer. Bishop and Varmus's pursuit of the *src* gene's origin had brought forth the notion of oncogene capture, and the understanding that healthy proto-oncogenes could convert to oncogenes through that capture. *Src* itself had nothing to do with the Abelson virus. But the concepts that had emerged from its study allowed Witte and others in the Baltimore lab to make sense of the giant protein inside cancerous mouse cells infected with the Abelson virus. The discovery led to an obvious question: Were tyrosine kinases some kind of universal mechanism for inducing cancer?

And that wasn't the only question at hand. The researchers also wanted to learn more about the gene behind the tyrosine kinase. Because they had been able to create a noncancerous strain of Abelson virus, Witte and Rosenberg suspected that a single gene was causing the cancer. But what was the gene like? Why did it lead to cancer? How had it gotten into the Abelson virus?

Just as Rosenberg and Witte had arrived at the right time with the right talent, so did Steve Goff, who came to the Baltimore lab in 1978 with expertise in cloning genes. Goff had read about recombinant DNA while he was an undergraduate at Amherst College and knew right away that using this technology was what he wanted to do. "Cloning allows you to do things with . . . viruses that you could once only dream of doing," said Goff, now at Columbia University. Today, the research techniques made possible by cloning—altering DNA, engineering mutations, combining DNA from two different sources—are as fundamental to science as letters are to reading. But in Goff's eager postdoc days, these techniques were entirely new, and their advent was perfectly timed for unraveling the inner workings of the Abelson virus.

Goff knew the history of the Abelson virus when he began working on it in the Baltimore lab. He knew that it had been derived from the Moloney virus, which caused a type of cancer called T-cell leukemia. He knew that the Moloney virus would cause this leukemia in nearly every infected mouse, but that the cancer might take months to develop. He knew that when Herb Abelson had ablated the thymus in a bunch of mice and then infected them with the Moloney virus, some of them developed B-cell leukemia. He knew that the virus extracted from the B-cell tumor would cause more B-cell tumors, even in mice with an intact thymus. He knew that the B-cell tumors arose rapidly—not in months, but in days or weeks. "In a month, the mice were all dead," said Goff. "So this was a really acute transforming virus." He knew that this new virus—Abelson virus—was made up of the Moloney virus plus some unknown element. But he didn't know what that unknown element could be; no one did.

Goff could use his recombinant DNA skills to find the added element in the Abelson virus. The place to start, though, was with the Moloney virus. Because its DNA was more abundant, and the lab already had probes for investigating its genome, Goff could more easily clone this virus. Afterward, he could compare the Moloney clone with the Abelson clone and zero in on where the two genomes differed.

Whatever element was present in the Abelson clone but not in the Moloney clone had to be the one he was searching for.

Goff was far from alone in these efforts; the field of scientific research was crowded with competitors. With other labs outside of MIT also trying to clone the Moloney virus genome, the race was on to be the first. The laboratory of Inder Verma, a former member of the Baltimore lab, was the first to clone pieces of the genome. Then Goff, together with MIT researchers Chuck Shoemaker and Eli Gilboa, extracted Moloney DNA directly from infected cells and created the infectious clone.

That preliminary task accomplished, he moved on to cloning the Abelson virus, as labs elsewhere focused on other genomes from viruses known to cause cancer transformation. As those other genomes were investigated, a pattern began to emerge: The viruses had always acquired some cellular gene. Study after study confirmed the "cellular origin" of proto-oncogenes, as Bishop and Varmus had declared. As the evidence accrued, the notion of cancer as a genetic disease began to sink deeper into the minds of researchers. When Goff and the others dived into cloning Abelson, they were pretty certain about what they would find: a viral genome with a new gene.

They were right. As soon as the genome was cloned, they could see the difference between Abelson and its precursor. "The new one is shorter than Moloney," Goff observed. "Pieces of Moloney are missing." That was interesting: The genome of the Abelson virus had some but not all of the genome of the Moloney virus.

There was one more observation. "There's unknown stuff also present," Goff noted. The Abelson virus contained genes that were not present in the Moloney virus. As Goff, Baltimore, Rosenberg, and Witte stared at the cloning results, wondering what this mystery stuff could be, Bishop and Varmus's proof that the RSV cancer transformation gene had been stolen from the host cell was echoing in their minds. "So we said, this has got to come from the mouse," said Goff.

Goff and others in the Baltimore lab removed the portion of the Abelson virus genome that was different from the Moloney virus genome. The separated fragment could then be used as a probe to find its match in the normal mouse genome. If they were right, if the extra gene in the Abelson virus had come from the mouse, then their probe would find it in the mouse genome.

What they found was remarkable: The mice did indeed have this gene, now called *abl*, short for Abelson. But the mouse's normal version—the proto-oncogene—was much bigger. The virus had a condensed version of the gene. Parts of the gene that did not pertain directly to the code for its product had been left out. Somehow, the virus had managed to acquire only the most essential information when it lifted the gene from the host cell. This made sense to Goff, who knew viruses had only very limited space for such codes. "Most genes are way too big to fit on a retrovirus," said Goff. "[So] the way it works is, the retroviruses just carry the message."

And they knew what the message was. Owen Witte had already uncovered that piece of information: The *abl* oncogene encoded a kinase.

How the virus had acquired that oncogene was now clear. When Herb Abelson injected the T-cell leukemia-causing Moloney virus into mice from which he'd removed the thymus, the virus just happened to scoop up a gene from the mice. That scooped-up *abl* gene just happened to be a proto-oncogene that, once it became part of the viral genome, turned into a full-on lethal oncogene. When this new virus, Abelson, infected mice, it caused a rapidly destructive B-cell leukemia. This was oncogene capture, just as Bishop and Varmus had described. Now, the mechanism by which the oncogene caused cancer was clear. The *abl* gene encoded a kinase, the activator, capable of setting in motion all manner of life-giving or -taking activities within. And this wasn't just any kinase; it was a tyrosine kinase.

WHILE GOFF WAS isolating the culprit oncogene and finding its normal counterpart in mice, Witte traded his New England winters for California sun, accepting a position at UCLA, where he continued to study the Gag/Abl kinase. He fiddled with the Abelson virus genome to make mutants that lacked the *abl* gene. These viral creations, Witte showed, did not induce cancer in mice, further confirming the central role of the Abl portion of the mutant kinase in cancer transformation.

Then, late one night in 1983, Witte, who was now running his own lab, got a call from one of his students, who had noticed something strange. The student had been doing antibody experiments just like the one Witte had used to fish out Gag, the protein to which Abl had been attached. By now, Witte had made an antibody that specifically recognized Gag/Abl as its antigen. The student had been adding the new antibody to various groups of cells to see how they reacted with it. If the antibody gravitated toward a protein in the cells, then those cells, he knew, contained either Gag, Abl, or Gag/Abl.

That night, he noticed that one group of cells had a very large protein that was reacting strongly with the antibody. Just as Witte had seen the Gag antibody do in cells from Abelson-infected mice, the Abl antibody headed straight for this unknown protein, taking it for a foreigner and attacking immediately.

The instant recognition surprised the student. He knew the cells had not come from a mouse infected with the Abelson virus. What kind of cells were these? Where had they come from? The antibody to Gag/Abl had been created to study a virus known to cause cancer in mice. Why was it so active in cells that hadn't come from an infected mouse?

He and Witte looked up the source of the cells and discovered that they had come from a patient with CML. That was a shock. Tyrosine kinases weren't linked to any human cancers. In fact, no causative link between genetic mutations and cancer had ever been made.

An idea began taking shape in Witte's mind. Why would an antibody to the Gag/Abl kinase react so strongly with human CML cells? Witte turned the matter over. Gag/Abl was a mutant protein encoded by a mutant gene that triggered B-cell leukemia in mice infected with the Abelson virus, itself an oddity that had arisen spontaneously in an NIH laboratory. What possible link could there be between Gag/Abl and CML? Oncogene research had not yet planted any stakes in the field of human cancer. Witte knew about the Philadelphia chromosome, but the idea that this abnormality could be connected to Abl was still out of reach. The thread of connection was so fragile and tentative, Witte was barely aware of the notion now taking form in his mind.

He knew that it was the addition of Abl that mattered. Goff's clones had proven that this gene was the difference between Moloney and Abelson. In becoming Abelson, the virus had lifted the *abl* gene from the mouse. Once in the virus, *abl* became oncogenic, triggering B-cell leukemia in the mice.

Now Witte began to wonder: Was there an Abl kinase present in human CML cells? Is that what the Gag/Abl antibody was reacting with in the cells his student had called him about?

Was Abl involved in human CML?

12

SPELLING OUT THE TRANSLOCATION

Witte and his growing crew of postdocs wasted no time in tearing into the mysterious CML protein. Which protein was reacting with the Gag/Abl antibody? What kind of protein was it? Was it involved in driving CML in people, as Gag/Abl was behind the B-cell leukemia of the Abelson virus?

In a completely separate pursuit, the Philadelphia chromosome was also coming under scrutiny. Rowley's 1973 discovery of the translocation had not fallen on deaf ears; it just happened that those ears were very far away from her Chicago lab. For years, two Dutch scientists—Nora Heisterkamp and John Groffen—had wanted to clone chromosomes. They weren't thinking of CML in particular, but they were aware of the increasingly recognized role of oncogenes in cancer, and wanted to dive into the work.

The development of cloning had engendered equal parts fear and hope around the world. Many federal governments restricted use of recombinant DNA—aka cloning—out of concern for what laboratory experiments could unleash. In this view, cloning was a biohazard disaster waiting to happen. It would bring new viruses that were more dangerous than the sum of their parts, animals with bizarre new traits, and whatever other unnatural catastrophes worry could conjure. But the promise of using cloning to understand human diseases was incredibly tantalizing.

At the time, an increasing number of diseases were being linked to genetic abnormalities. Hereditary disorders like hemophilia, in which blood clotting is impaired, and Klinefelter's syndrome, which affects male development, had long been traced to variations in the X chromosome, one of the two chromosomes specific to sex. The 1959 discovery of the extra copy of chromosome 21 (known as trisomy 21) in people with Down syndrome fueled the already growing interest in the link between genes and disease, and brought the emerging field of cytogenetics into greater prominence. In the 1970s, Tay-Sachs disease, a well-known inherited disease most common in the Jewish population, was connected to mutations on chromosome 15.

Cloning portions of DNA using recombinant technologies would allow geneticists to explore those links more closely and to identify others. Although treatments targeting such genetic abnormalities—the prominent thrust of personalized medicine in the twenty-first century—were just hovering on the edges of anyone's imagination, exploring the genetic underpinnings of human health and disease was considered essential scientific research. In 1976, the United States provided its first guidelines on the use of recombinant DNA, opening the floodgates ever so slightly. Other countries gradually followed suit.

In 1978, Groffen went to England to learn cloning techniques with another Dutch geneticist who'd left his home country, where recombinant DNA was still forbidden, to pursue his research elsewhere. Eventually, Groffen and Heisterkamp, who'd been postdocs in the same Netherlands lab, landed at the National Cancer Institute. Groffen's assignment was to create a library of genes, noting their location on the human genome. The Baltimore lab's work had been with mice, but by now geneticists knew that a gene in a mouse would almost certainly have a human genome counterpart. Having read papers about the *abl* oncogene from Baltimore's lab, Groffen was interested to see if he could figure out the exact location of the normal *abl* gene on the human genome.

Just as Bishop and Varmus had so laboriously done with *src* before the invention of recombinant DNA, Groffen created an *abl* probe, a single strand of DNA containing only this gene. With that isolated sequence in hand, he could search for the exact location of the matching sequence within the human genome. Research at the Baltimore lab had found the fusion gene, *gag/abl*, responsible for the cancer-inducing capacity of the Abelson virus, and the fusion kinase Gag/Abl, encoded by the gene. At Owen Witte's lab in California, signs of a similar protein in human CML cells had been spotted. The *gag* gene belonged exclusively to viruses, but *abl* was a mouse gene, and most likely the same gene existed in the human genome. But where among the entire array of human chromosomes was *abl* located? The work took months, but finally, in June 1982, he had the answer. The human *abl* gene was located on chromosome 9.

Over an evening glass of wine with another Dutch geneticist visiting the United States that summer, Heisterkamp and Groffen began wondering about whether *abl* might have some connection to CML. After all, chromosome 9, they knew, was involved in the Philadelphia chromosome translocation. Nowell and Hungerford had first spotted the abnormally small version of chromosome 22, and Janet Rowley had recognized this change to be part of a swap with genetic material on chromosome 9. More than 95 percent of people with CML had that genetic mutation. Could *abl* be involved in that translocation? It was just a musing. Each chromosome contains hundreds or even thousands of genes, so chances were slim that the translocated bit included *abl*. But by the time drinks were over, the notion had grown roots.

As it happened, that visiting geneticist had a brother, also a geneticist, who had just been given the task of cloning the Philadelphia chromosome breakpoint, that is, the exact spot in the chromosome where the breaks occurred in the swap. So, Heisterkamp and Groffen sent their *abl* probe—the mechanism for detecting *abl* in a genome—to the brother, Gerard Grosveld, back in Holland. Because the probe contained the genetic sequence of *abl*, Grosveld could use it to check for a match in the mutant Philadelphia chromosome.

He applied the probe to cells from CML patients. Soon enough, he found the matching gene. But it wasn't on both copies of chromosome 9, as Heisterkamp and Groffen had found in normal cells. In one copy of chromosome 9, *abl* was there. But the other copy of chromosome 9—all chromosomes come in pairs, one inherited from each parent—didn't have *abl*. A portion was there, but *a large portion of it was missing.*

And that missing portion was on chromosome 22. The significance was immediately obvious: Grosveld knew that 22 was the other chromosome involved in the Philadelphia translocation. Groffen and Heisterkamp's hunch had been correct: *abl* was located on the piece of chromosome 9 that switched places with a piece of chromosome 22.

Heisterkamp and Groffen set to work mapping chromosome 22 to see what exactly was going on. When *abl* moved to chromosome 22, what gene did it end up next to? What proteins did those genes encode? They knew that a mutation that combined two genes could result in two proteins being lumped together; Witte's discovery of the Gag/Abl kinase, a fusion of two previously separate proteins, pointed to an underlying fusion of two previously separate genes. So checking out *abl*'s new neighbors made sense.

In 1984, they found the region on chromosome 22 where the break occurred when its tip translocated to chromosome 9. If the chromosome could be conceived to be a mile long, they had found the few inches in which the break always occurred. Of the nineteen CML patients whose DNA they studied, seventeen had a break at this same region of chromosome 22. In patient after patient, the break occurred at almost exactly the same point. They called it "breakpoint cluster region," or *bcr.* In unraveling the sequence they'd homed in on, Heisterkamp and Groffen could see that *bcr* was a distinct gene. It had the telltale patterns in the A, C, T, G base pairings that mark the beginning and end of an individual gene sequence within the long, coiled strand of DNA. They had no clue what *bcr* did.

Just then, a third researcher entered the story. In Israel, Eli Canaani began investigating the protein product of *abl*—in genetics parlance, how *abl* was expressed in cells—in CML patients. As with so many scientific pursuits that catch fire, research on the *abl* gene and Abl kinase was lighting up a global map as lab groups around the world followed the onslaught of reports in scientific journals and began brandishing their own scientific swords as they joined the crusade. Each lab strived for the next significant breakthrough, fomenting competition but also speeding up the story.

Canaani wanted to see if the RNA made by this gene was different from the RNA made by the normal *abl* proto-oncogene that existed in normal, non-CML cells. Was the instruction encoded by *abl* in healthy people different from the instruction encoded by the version of *abl* present in people with CML?

It was. The RNA in the CML cells that Canaani was investigating was not present in normal human cells, those that lacked the Philadelphia chromosome translocation. In CML cells, the RNA was present in copious amounts. Here was the third piece of crucial evidence that *abl* was directly involved in the development of CML.

Suddenly, the story had taken on a life of its own, as if it had somehow amassed enough material to have a gravitational pull. Facts began coming together in rapid succession. In 1984, Witte and his group reported that the CML protein, the strangely large one that had been so quickly engulfed by the antibody with which his student had been experimenting, was a mutant form of the Abl kinase. They weighed the mutant protein and found out that it was much heavier than the normal Abl kinase, suggesting that something had been added to it.

The discovery mirrored exactly what Witte had found in mice infected with the Abelson virus. When he had extracted a protein with the Gag antibody from mice infected with the Abelson virus, he'd found that the Gag protein had something else attached to it. That something else had been named Abl, the product of the *abl* oncogene. Gag and Abl had fused together to form the supersized protein Witte had called Gag/Abl. Now, he discovered that the Abl protein, a tyrosine kinase, in CML cells had also fused to another protein, just as it had done in the mice. What that other protein was he didn't know.

At nearly the same moment, Heisterkamp and Groffen completed their map of chromosome 9. They knew a piece of the *abl* gene was broken off and moved to chromosome 22. And they had isolated the exact region on chromosome 22 where that break occurred, within a gene subsequently named *bcr*. Now they showed that the broken-off piece of *bcr* from chromosome 22 was next to what remained of the *abl* gene back on chromosome 9. In CML cells that had the Philadelphia chromosome, *abl* was fused to *bcr*. The two previously distinct genetic sequences had melded together, to become *bcr/abl*.

This change, they knew, had to be responsible for the cancer-causing tendencies of *abl*. The only difference between normal cells and CML cells was that in the former, *bcr* and *abl* were separate and that in the latter, *bcr* and *abl* were fused. And that fusion turned the once-harmless *abl* into an oncogene. Instead of the normal Abl kinase of healthy cells, *bcr/abl* expressed an abnormal kinase that induced CML. That fusion protein had to be the cause of CML.

There was one final question that no one had yet answered. How did Bcr/Abl cause cancer? Everyone knew it was a tyrosine kinase. They knew, from years of research at the Cohen lab in Dundee and elsewhere, that kinases set events—such as the production of blood cells—in motion by taking phosphates from ATP and placing them on proteins sitting at the top of the pathways responsible for those events. Adding Bcr/Abl to other research, tyrosine kinases appeared to be involved in many cancers, leading many researchers to suspect them as a general mechanism by which cancer operated. But what was the difference between normal tyrosine kinases and those involved in cancer?

A rapid succession of research spelled out exactly what was happening. The main clue came from a leukemia cell line known as K562. These cells were derived from a CML patient in blast crisis, intentionally created for laboratory research. In 1984, Witte and his lab reported that a kinase in those cells was continually active. That is, instead of transporting phosphates to proteins only intermittently, the kinase was doing it all the time. Two years later, the Baltimore lab made the connection that the kinase in that cell line was the same one encoded by the *bcr/abl* gene in human CML. That link sealed the conclusion: The Bcr/Abl tyrosine kinase caused CML because it was deregulated. It never stopped powering up the protein responsible for launching the pathway leading to white blood cell production. The kinase was, in scientific terms, constitutively active. In more accessible language: The Bcr/Abl tyrosine kinase was out of control.

13

"THAT WORD IS ONCOGENES"

The story was now clear. Finally, the work done by Nowell, Hungerford, Abelson, Rowley, Erikson, Hunter, Rosenberg, Witte, Goff, Heisterkamp, Groffen, Grosveld, Canaani, and a score of others during those twenty-five years had come together. Many of these people knew of each other over the years, but just as often they were strangers separated by years, geography, and the unawareness that their pursuits were connected. None of them envisioned how their strands of research would finally weave together.

The Philadelphia chromosome was a translocation that brought together a piece of *bcr* and a piece of *abl.* This fusion gene created a fusion protein. Normally, *abl* encoded a tyrosine kinase. The fusion protein was a tyrosine kinase with heightened activity. A bit of Bcr put the Abl kinase into a perpetual "on" setting. In that abnormal mode, Abl would not stop phosphorylating proteins. In a never-ending loop, the kinase plucked phosphates and stuck them onto the tyrosine key hooks on proteins inside white blood cells. This continuous phosphorylation triggered a signaling pathway that resulted in the excess production of these cells. That addition of just a portion of the *bcr* gene—a tiny scrap of DNA sequence—turned the Abl kinase into a killing machine.

The unfolding of events was uncanny. Herb Abelson's experiments had led to the discovery of a spontaneously arising virus. If the thymus-annihilating experiments were repeated today, the virus wouldn't necessarily appear. ("In later years we talked about repeating the whole experiment, but never got around to it," Abelson said. "Biology is testy and fickle.") At MIT, the Baltimore lab discovered that the virus had engulfed a proto-oncogene that, once integrated into the viral genome, became a leukemia-causing oncogene in mice. That understanding had been made possible by work on the opposite coast when Bishop and Varmus identified the cellular origin of the *src* oncogene. Their work, derived from decades of research on a virus extracted from a chicken by Peyton Rous in 1912, had directly informed the study of the cancer-causing Abelson virus. Then Witte identified the cancer-inducing protein encoded by *abl* as a tyrosine kinase, a fact that was the scientific descendant of Ray Erikson and Marc Collett's discovery that *src* encoded a kinase and of Tony Hunter's revelation that this kinase phosphorylated a previously ignored amino acid called tyrosine.

The fact that the culprit gene in the Abelson virus was the culprit gene in this human leukemia came as a shock even to the most seasoned researchers behind the work. That the fusion gene and fusion protein Witte isolated in Abelson-infected mouse cells would have an exact parallel in the fusion gene and protein isolated in human CML cells was completely unforeseen. And then to find, as Heisterkamp and Groffen had, that the human version of *abl* was on chromosome 9 and moved to chromosome 22 in the Philadelphia abnormality—the story was almost too good to believe.

The pursuit of the *src* oncogene's origin and that of its protein product had begun as distinct efforts to unravel the mechanism behind virally caused cancer. These strands of research had been completely separate from the Philadelphia chromosome and from the Abelson virus. Gradually, unknowingly, they had become inextricably woven together.

Recombinant DNA and chromosome banding techniques had come along at just the right time. Herb Abelson ended up just upstairs from the Baltimore lab. Witte, Rosenberg, and Goff brought the right skills to the right place at the right time. And the research had all been careful and airtight.

The way the Philadelphia chromosome led to CML was a story that unfolded like a hundred painters applying brushes to a canvas at some time or another over twenty-five years, driven only by curiosity and, sometimes, a vague hope that their work might eventually be relevant to human cancer. There'd been no final picture in mind and no awareness that they were even painting something together. And yet there it was. A scientific masterpiece.

In August 1984, Nick Lydon, the researcher who'd been so captivated by kinases at the University of Dundee, was enjoying a warm summer in Dardilly, a small town in southern France just outside Lyons. He'd left Dundee for Paris in 1982 to work for the pharmaceutical company Schering-Plough, which had just opened a new lab in the quiet countryside. On a lunch break, he picked up that month's issue of *Scientific American* and began to read an article by Tony Hunter, "The Proteins of Oncogenes." In it, Hunter, the scientist who'd found the first tyrosine-aimed kinase, described all that had been discovered about kinases. Again and again, proto-oncogenes were found to encode kinases, and as the gene transformed from normal to mutant, so did the kinase move from healthy to cancer inducing. A shift in the gene transformed the kinase into its haywire state, continually phosphorylating proteins and thus continually sending the signal for the cell to reproduce. It was, as Witte and others had suspected, a widespread mechanism for cancer transformation.

One final proof remained to confirm without a doubt that the mutated Bcr/Abl kinase—and that alone—was sufficient to cause CML. That step was necessary in order to consider the link an established fact. Even with all of the work that had been accomplished, the connection between Bcr/Abl and CML could still only be called an association. Someone had to demonstrate that this abnormal protein alone was responsible for the cancer. When that happened, the abnormal kinase could be called the *cause* of CML.

But for Lydon, that last proof was a formality that would come when he was well on his way toward his new goal. By 1984, the weight of the work already done was enough to convince him of the wayward kinase's involvement in CML. Adding all that up with Hunter's magazine summary solidified Lydon's view that the kinase, as a general mechanism, was a driving factor in the development and progression of cancer. This enzyme—the one that had so piqued his curiosity during his postdoc years, the protein product of the first oncogene ever identified, the substance connected to nearly every oncogene identified to date—compelled Lydon toward a singular mission: creating a drug that would block the kinase and thereby stop the growth of cancer.

At around that same time, Brian Druker was having, as he puts it, his "plastics moment," referring to the famous scene of *The Graduate* when a family friend tells Dustin Hoffman's character that his future lay in plastics. Having finished medical school near the top of his class, he'd gone to Washington University for a residency in internal medicine at Barnes Hospital, one of the most rigorous training programs in the country. These were still the years before limits were put on resident shifts, and Druker was working 100-hour weeks, on call for multiple nights each week. It was another proving ground, one in which the number of admissions handled was a badge of honor, and asking for help was a sign of weakness. Spare time was nonexistent; he and his friends (a group that included Lawrence Piro, who would later become famous as Farrah Fawcett's oncologist, and Brian Kobilka, who won the 2012 Nobel Prize in Chemistry) would frequently scrape together pennies for after-work pizzas, because there'd been time to stop at a bank.

But Druker had an advantage: his inexplicable interest in cancer. He'd already pushed himself to handle particularly difficult rotations during medical school, putting him at relatively greater ease with his new environment. "I was comfortable taking care of the sickest of the sick," he recalled later. And he flourished, becoming the go-to person for questions, taking on extra hours. "I loved taking care of patients," he said.

It was during this time that Druker had his first exposure to bone marrow transplants, a dangerous procedure that for many years was the only hope for curing CML. At the time, high doses of chemotherapy, followed by a bone marrow transplant, was a treatment option for colon cancer, melanoma, breast cancer—basically any metastatic, terminal cancer. "In those days, it was essentially a one-way ticket out," said Druker. "The issue was: Do you torture somebody before they die, or do you just resign to the fact that they are going to die?" Bone marrow would be removed from the patient, who then received drugs followed by a reinfusion of their marrow. But patients frequently ended up in the intensive care unit before the marrow could recover. "Almost nobody made it out of the [ICU]," says Druker. "It was just this incredibly frustrating experience."

If that horror did anything for Druker's medical career, it was to strengthen it. He would sit by patients' bedsides long after his shift had ended, and he would administer treatments himself rather than hand a distressed patient over to someone unfamiliar. He was there for the first patient at Barnes to ever be given human insulin, a landmark breakthrough in the treatment of diabetes, especially for people who were allergic to pork or beef insulin, the only prior options. That was the first time he witnessed a man being brought back to life from the brink of death.

But at the end of those three years, it was time for Druker to finally settle on a specialty, a decision he'd been resisting for more than ten years. Unable to deny his wish to focus on cancer any longer, he finally mustered the courage to speak it out loud.

"I think I want to do cancer," he told David Kipnis, the chair of medicine at Washington University at the time.

"'I'm going to tell you one word. Are you listening?'" Druker recalled Kipnis replying, a memory colored by *The Graduate*'s famous dialogue. "'That word is oncogenes.'"

Aside from the cloistered world of his residency, Druker knew nothing about what was happening in the world of cancer research. He had never heard of the Philadelphia chromosome. He knew nothing about the Abelson virus, the Bcr/Abl tyrosine kinase, or any of the work that had been going on during the past two-plus decades. "All I knew was that I was sleep deprived and trying to take good care of patients," Druker said later. But he listened to Kipnis. "He understood where the future of cancer research was going. This is going to be where the discoveries will happen, and I needed to pay attention to that," Druker said. "And he was absolutely right."

Despite the intensity of his residency, Druker never forgot about his interest in lab research. So when it came time to decide exactly how to pursue his finally declared interest in cancer, and now armed with the forcefully delivered insights from Kipnis, Druker knew where he wanted to go: Dana-Farber Cancer Institute. For someone who wanted to research cancer in the lab without losing touch with patient care, Dana-Farber was the perfect match. His fellowship training began there in 1984. He arrived in Boston at virtually the same time that Lydon was reading Tony Hunter's *Scientific American* declaration about the role of kinases in cancer and just a few years before the two would meet.

PART 2

Rational Design

1983–1998

Although logic and evidence had become the pillars of modern medicine, cancer treatments were mostly the desperate products of trial and error. Drugs were not made to target the cause of a malignancy, and no methods existed for creating medications that would do so.

Following the discovery of the Philadelphia chromosome, blood from every patient diagnosed with CML was tested for the genetic abnormality as a matter of routine. But the information was meaningless to the care and survival of people with CML. No one knew why almost every patient with CML had the Philadelphia chromosome. Even when the connection between the genetic mutation and the cancer was understood, the science was irrelevant to the patients harboring the abnormality.

14

BECOMING A DOCTOR, AND THEN A SCIENTIST

By the time he joined Dana-Farber in 1983, Brian Druker had already been thrown around and tossed ashore by the rough waves of medical training. He had obtained his medical license and had come out the other side of a rigorous training in internal medicine. As he looked back, he saw how those years had been stamped with his low-key ways. He'd always been a successful student, but his mellow mannerism didn't attract the attention that the type-A students had always seemed to garner, sitting at the front of the classroom and raising their hands, and he'd never been good at playing the game. (When a med school interviewer had asked him what scientific prizes he dreamed of winning, he answered with the names of actual prizes; the thought of replying, "Just doing the work is enough for me" hadn't even occurred to him.) Those traits, combined with his reticence about pursuing medicine in the first place, left him under the radar.

With each step, he'd had to find his footing and figure out where he stood in the pecking order. Would he be among the top students at medical school? Would he be among the best doctors at Barnes Hospital? He'd never minded the fact that his ambitions and interests seemed to run counter to those of his colleagues. But he knew he was smart, and a part of him craved recognition for that fact, a legacy of an upbringing where academic success was prized above all by his parents. By the end of his residency years, he'd clearly proved himself to be an excellent physician. He was the one always willing to take on the most difficult patients. He'd excelled in his medical knowledge, and his understated personality made him a calm and reassuring presence at a hospital bedside. But now here he was at Dana-Farber with doctors who'd been in the Harvard system since their freshman year in college. The old insecurities came flooding back.

He was assigned about eighty patients during his first fellowship year, now with no senior resident looking over his shoulder, as there had been at Barnes. Though he'd cared for some cancer patients in those years, he still had only very little official training in oncology. By his own admission, he was basically clueless about how to treat cancer. But the stigma against asking for help was pervasive in medical training, and he was loath to show signs of weakness. "How much am I willing to admit that I don't know versus the fact that I'm practically drowning?" Druker said, recalling his state of mind that first year.

Cancer proved to be an efficient teacher. A senior colleague insisted that Druker ask for help when he needed it but added that after three months he'd be so familiar with the treatments he wouldn't need to ask anymore. "That was absolutely true," he said. He immersed himself in reading about cancer drugs, learning what tests to use to tell if a treatment was working, and diagnosing each specific type of cancer. His patient roster included Dana-Farber trustees and inner-city drug addicts.

Then came John Bullitt, who taught English at Harvard University and had been diagnosed with lung cancer after decades of chain smoking. The gruff professor took a shine to Druker right away. On the day that Druker explained the diagnosis and treatment to Bullitt, the attending physician—who had to sign off on the work of first-year fellows like Druker—was one renowned both for his breast cancer expertise and his huge ego. He introduced himself pompously to Bullitt and talked casually about his treatment, which he really knew nothing about, and then left abruptly. Bullitt saw right through his act. As soon as the door closed, he turned to Druker and said, "He is full of shit." Bullitt had recognized the doctor in Druker, and hadn't needed reassurance from another. In that moment Druker finally allowed himself to become a doctor. "I'm okay, I'm a doctor," Druker remembers realizing. "I'm going to be an oncologist." According to the medical license authorities, he'd already become a doctor. Now he'd become one on his own authority.

That first year had more to teach Druker. If he was to come to terms with being a cancer doctor, then he had to learn to cope with death. Most of his patients would die, if not shortly after coming under his care, then in the easily foreseeable future. That was what happened with cancer: Almost everyone died from the disease. So he learned to temper his expectations about what he could do, reasoning that if he could add a few months of quality time, that was a good enough goal. He learned that unless his patient had Hodgkin's disease (a type of blood cancer) or testicular cancer, the only curable cancers at the time, he or she would die. "You have to face that fact, and you have to protect yourself a little bit," said Druker. He learned to talk to the families of his dying patients, especially after the harshly worded letter one family wrote to Dana-Farber after Druker sided with an end-stage patient who had decided against further treatment. He took up the institute's tradition of writing to the families whose loved ones had died under his care, and he attended funerals of the patients with whom he'd been closest. Dana-Farber's practice was to allow the fellows to continue treating their choice of forty or so patients as they headed off to their laboratory training after the first year of their fellowship. By the end of that second year, Druker was left with about five patients, and by his third, perhaps one or two, as each one succumbed to cancer.

In 1985, in the thick of his first year, his schedule still filled with patient care, it was time to choose a lab. "In sort of typical fashion, I had no clue," says Druker. He was engraved with David Kipnis's advice to pursue oncogenes, but he had no idea how to choose a lab where he could do that. He heard from a friend about Tom Roberts, an up-and-coming young doctor at Dana-Farber who was studying oncogenes. "So I went and talked to Tom," says Druker, "and he was talking this language I didn't understand."

That foreign language was, of course, the language of laboratory research. Roberts was a PhD scientist who'd followed a straightforward path to lab research, unlike Druker. Years earlier, Druker's lab research had brought him on an unconventional course toward medical school. Now he was circling back to the lab by way of medical school, another path less taken. So when Druker asked if he could join the lab, Roberts wasn't sure. His response—"I've never taken a medical oncology fellow before"—exemplified the chasm between the laboratory and the clinic that has pervaded medicine until only recently. Druker insisted that he could get his own funding, that it wouldn't cost Roberts a dime to have him there, and so Roberts agreed.

From there, Druker plunged straight down to the bottom of the ladder. Here he was, a board-certified internal medicine specialist coming off a year of intense oncology training, and he knew absolutely nothing about laboratory research. He struggled to get through articles on the latest oncogenes research. Every day he was surrounded by people who seemed to know exactly what they were doing, and he was completely clueless. Taking care of dying lung cancer patients looked simple from his new vantage point.

To be fair, the world of oncogenes had become vast and complex since Bishop and Varmus had first announced the cellular origin of the *src* oncogene. Their discovery still stands as a landmark moment in the history of cancer research. Now, the study of these cancer-inducing genes had taken on a life of its own, separate from the study of cancer-causing viruses. In the late 1970s and early 1980s, Robert Weinberg, yet another eminent scientist just down the hall from David Baltimore's lab at MIT, along with Michael Wigler of Cold Spring Harbor Laboratory and Mariano Barbacid of the National Cancer Institute, independently reported genes they'd found in the DNA of tumor cells that, when extracted from those tumor cells, could transform normal cells into cancer cells in culture. These oncogenes had nothing to do with oncogenes found in retroviruses like the Rous sarcoma virus and the Abelson virus, yet in many cases they were proving to be extraordinarily similar.

Before 1975, cancer could not be understood in terms of genetics because cells couldn't be studied at that level. From 1975 to 1985, the landscape of cancer had become more detailed. It was like moving from the broad strokes of Matisse to the pointillism of Seurat. As researchers like Weinberg were uncovering how genes transformed from proto-onco to onco—through mutations caused by carcinogens, for example—the world at large was beginning to see cancer through gene-tinted glasses.

A cancer doctor now becoming a cancer scientist in the mid-1980s, Druker personified that shift and its accompanying brain-scrambling struggles. Not only was he without basic research skills, he was without any knowledge of the field. He had no idea what kinases were. He had no idea about the Abelson virus, or *src*, or chromosomal translocations. "Now there's a history to molecular biology and cancer research," said Druker. "At that time, there wasn't any history."

In the mid 1980s, many geneticists believed that there would ultimately be about 100 oncogenes found among the total complement of human genes. And, Druker and others were being told, genes with cancer-causing potential would probably all be ones that were highly conserved; that is, they'd been around for a long time. Highly conserved genes are the ones found in vast numbers of different species, indicating their importance throughout evolutionary history. It's like a chef who has used the first knife she bought for cooking school to make ever more complicated dishes; that knife is highly conserved.

Unlike Naomi Rosenberg and Herb Abelson, whose contributions to the Philadelphia chromosome story were directly related to the freedom they were given to pursue their ideas, Druker's research career began in the opposite trajectory. Clueless as he was about the laboratory, he needed a concrete project that would serve as an introductory course. Roberts assigned him to study polyomavirus, which causes tumors in rodents. It was the same virus that had led Tony Hunter to his discovery that tyrosine, a rare and unexplored amino acid, served as the binding site—the landing platform—for the phosphate delivered by the kinase. Roberts wanted Druker to lay out the cascade of events that led from infection to cancer transformation.

For scientists interested in oncogenes, polyoma had a significant advantage over many other cancer-inducing viruses. The revelation delivered by Bishop and Varmus about *src* was that it originated not in the virus but in healthy mammal cells. The oncogene version of *src*, the one that expressed an abnormal kinase, was just slightly different from the proto-oncogene version, the one that existed in healthy cells that expressed a normal kinase. Their similarity made them tricky candidates for biochemical experiments because it's hard for researchers to be sure which version they're dealing with at any given moment, like telling the difference between identical twins at first introduction. By contrast, the oncogene in polyomavirus doesn't have a normal, proto-oncogene counterpart. As an only child, it was a much easier biochemical system with which to work.

Druker proceeded with no clear idea of where the work would lead him. The world of cancer research was not wondering about the relevance of such work to human cancer. "Nobody cared about that," says Druker. "People didn't talk about translation to the clinic." Today, terms like "bench to bedside" and "translational research" signify the growing pressure for scientists to aim their work toward a practical result with some tangible benefit—for the microscopic world of research to have more immediate relevance to the macroscopic world of human suffering. But at that time, the lab and the hospital were distinct worlds, and scientists and doctors did not view themselves as aligned to a common goal. Clinical faculty like Druker who entered lab research were considered as good as gone, an attitude with some foundation. "You [had] people who went into the lab and did their research, and nothing ever came out," says Druker. "It was just this big, black hole that you'd send people into, and you'd just never see them again." As he began to fiddle with the microscopes and test tubes around him, Druker wondered if he was heading toward the same fate.

His job was to unravel the so-called signal transduction pathway by which polyomavirus caused cells to become cancerous. He would break down the hundreds of amino acids in the virus into blocks of fifty or fewer, to see which increasingly small group held the secret to the cancer transformation. If he mutated these ten amino acids, does the tumor still occur? Just as Sir Philip Cohen's pioneering research had done with insulin at Dundee, Druker was trying to piece together the stream of signals launched by the kinase, like a relay race in which the gene shoots the gun and the kinase is the first runner. Roberts and Druker were trying to trace the baton from the starting line to the finish line.

The work was grueling, essentially requiring Druker to make mutations of around 500 amino acids, one at a time. And the information the research yielded was slow in coming. "It got me a lot of training," said Druker, "but didn't get me a lot of actual publications." The discoveries he made turned out to be more internal, more personal development than cancer development. Over the course of the five years he spent on the project, he learned what it was to be an academic lab researcher. He'd become a doctor, on paper and in thought. Now, he was becoming a scientist.

Two years into the fellowship, Druker was working in the lab on a weekend, doing a routine preparation, cutting and pasting genes and trying to figure out whether they recombined correctly. The work had become automatic for him, leaving his mind free. On this particular day, he started questioning what he was even doing in the lab. Why had he gone to medical school? Why had he done the intense residency only to get entrenched in basic research? He still believed the lab work would lead him somewhere, but he missed seeing patients and worried about losing touch with medical care. "Someday I may want to do something that benefits patients," Druker remembers thinking. "But how can I do that if I've lost those skills?"

His discontent was well timed. A position was opening up at Nashoba Community Hospital, in Ayer, Massachusetts, a town not far from Boston. The job, medical director of the oncology clinic, would require just one day a week of patient care. The arrangement was ideal: Druker could keep his connection with patients and spend the other six days at the lab (taking days off was not something that occurred to him).

A nebulous goal began to form in his mind. He knew that his urge to continue treating patients was for a specific purpose. "Somebody might figure something out, and I might want to run a clinical trial," he said. "I wanted to be part of that." He was a good doctor who'd seen the inadequacy of cancer treatments time and time again. Now, for the first time, he began to articulate why he had ended up in the laboratory after medical school, and why that had driven him back to the clinic: If molecular biology discovered how cancer begins and progresses, he wanted to help convert that knowledge into something that would improve the lives of patients.

Following close on the heels of that decision came his first significant contribution to cancer research. In the late 1980s, a woman named Deborah Morrison, who had expertise in making monoclonal antibodies, joined the Roberts lab. She possessed the same skills that Owen Witte had brought to the Baltimore lab a decade earlier, creating antibodies to tease out proteins. Druker knew that if he were to run his own lab someday, he'd need to know how to make antibodies for specific proteins, and he took the opportunity to learn all that he could about it from this new researcher.

The best way to learn the technique was to just go ahead and do it. He decided to make one for phosphorylated tyrosine, using mice as the antibody-producing animal. Druker was aware that research was pointing to tyrosine phosphorylation as a general mechanism driving cancer. Tyrosine was the surprising amino acid to which the kinase encoded by *src* bound phosphate, the bit of energy that powered up the protein and appeared in other signaling pathways that started with a kinase and resulted in cancer. Witte had discovered that Bcr/Abl, the fusion protein that induced CML, was a tyrosine kinase, but Druker wasn't thinking solely about CML. He just knew that tyrosine was emerging as an important substance in cancer research.

By the late 1980s, he was also hearing murmurs about the possibility of creating drugs that could block the tyrosine kinase as a way to stop the progression of cancer. If a cancer resulted from the haywire activity of an abnormal tyrosine kinase—as seemed to be the case with Bcr/Abl—then could cancer be stopped by somehow turning off the kinase, by halting it in its tracks? Kill the kinase and you kill the cancer, the theory went. Around the world, a handful of experiments were showing that theory might just pan out.

The antibody Druker was trying to create could be an invaluable research tool, he knew, because it would show how much tyrosine had been phosphorylated. An antibody against phosphotyrosine—the shortened way of referring to phosphorylated tyrosine—would automatically latch onto the substance, as antibodies do to foreign invaders. That attack would draw the target protein away from the rest of a cell's contents, enabling it to be measured. If a drug inhibited the tyrosine kinase, then the amount of phosphotyrosine in drug-treated cells would be less than in untreated cells. And if the amount indeed decreased in a drug-treated sample, then the drug was doing something—at the very least, it was blocking the kinase from its phosphate-carrying work. In theory, that inhibition would prevent the protein at the top of the cancer-causing cascade from being switched on. If the incessant signal from the kinase to produce white blood cells never came, then cancer wouldn't develop.

Those thoughts, however, were vague and distant in Druker's mind as he struggled over two years to make the antibody. His goal was only to learn how to make the antibody. But although the two years of efforts taught him how to do it, he was having no luck getting a final product: Every attempt was a failure. Finally, the lab contracted the work out, and it soon had its antibody, called 4G10, to phosphorylated tyrosine. From there, Druker could grow the clone, purify it, and experiment with it. But the primary mission had already been accomplished: He had acquired the skill. "It may be useful someday," he thought at the time. As it turned out, the work he had done was about to come in very handy, but not in the way he anticipated.

There was one other unexpected consequence of his work on 4G10. In 1988, the other lab member working on the project with him introduced Druker to her roommate, Barbara. She and Druker began dating, and two years later they were married.

15

TURNING A PROTEIN INTO A DRUG TARGET

In the early 1980s, back when Druker was just starting to learn basic laboratory research and nursing the shrunken ego that came with returning to the bottom of the ladder, Nick Lydon, the man who'd skulked the halls outside Phil Cohen's lab at Dundee to gather whatever he could about kinases, was already gripped by the possibility of creating drugs to block the cancer-causing enzymes.

And he wasn't the only one. His boss, Alex Matter, had also lit on the idea. Matter had a brief clinical career before entering pharmaceutical research. Though Matter's work as a bedside oncologist had been brief, it was enough to leave him haunted about the state of cancer treatment. "[It] was an absolutely earth-shattering experience," said Matter, who'd been devastated by the death of one of his patients, a young mother with ovarian cancer who'd left behind three children. His degree of helplessness was equally upsetting. "There was nothing that I could actually do about it," he said. He decided to move into industry. A job focused on making better drugs seemed more worthwhile than treating patients with the available medications.

In the late 1970s, Matter took a job at Roche Pharmaceuticals, where a colleague talked about new ideas that were starting to enter the drug development arena, including the possibility of stimulating the immune system as a way to eradicate cancer, or forcing cancer cells to age and die off using substances derived from vitamin A. Matter had some success developing drugs at Roche, but never for any serious cancers. The breakthrough he'd been hoping for eluded him.

A few years after joining Roche, Matter moved to Schering-Plough, where Lydon had gone after Dundee, as head of oncology drug development. The company had just acquired a drug called interferon, a synthetic version of a substance that occurs naturally in the body. Interferon stimulates the immune system, and for Matter it was the first exciting drug of his career. In some instances, the drug could eradicate cancer. Interferon turned out not to be the panacea that many researchers had predicted it would be, but it did work in several different types of cancer, including melanoma, a type of leukemia called hairy cell leukemia, and also CML. (Interferon's greatest benefit turned out to be in the treatment of viral infections; it was the standard treatment for hepatitis C until only recently.)

A year or two after Matter and Lydon began working together, Schering opened up a laboratory in Dardilly, a rural village in the south of France. The lab, with about sixty-five people, had been created specifically to study interferon and other so-called immunostimulants. Matter was asked to serve as the director, and Lydon went with him. But shifting finances resulting from changes in the political dynamics of France at the time, coupled with Schering-Plough's simultaneous acquisition of the California-based company Dinex, which had been focusing on molecular immunology, dimmed the promise of the Dardilly lab. In a very short time it became obsolete, shrinking to twenty employees. At the same time, Matter was getting fed up with the bureaucracy of the sprawling company. He had supervisors in Paris, New Jersey, and San Diego, and he constantly had to make the rounds to keep them informed and appeased. Eventually, the situation became explosive. In the early 1980s, Matter left the company on unfriendly terms.

He accepted a job at Ciba-Geigy, a pharmaceutical giant situated along the Rhine River in Basel, Switzerland. Despite the administrative difficulties he'd had at Schering, the years there had left him inspired about creating pharmaceuticals for cancer. He'd been following the studies of oncogenes and kinases avidly, and he wanted to turn those discoveries into drugs that would improve the outlook for cancer patients. It was the same dream that would bring Druker to Nashoba Community Hospital just a couple of years later.

When Matter was still at Schering, he and Lydon spoke frequently about kinases, how they seemed like the perfect drug target. By the time Matter arrived at Ciba-Geigy, links between oncogenes and kinases were cropping up in several cancer research labs. It was far more than *src* at this point. A cancer-causing protein known as v-erbB turned out to be related to a kinase called epidermal growth factor receptor, or EGFR. Another, v-sis, was connected to a kinase called platelet-derived growth factor receptor, or PDGFR. In the late 1970s, a Japanese researcher named Yasutomi Nishizuka discovered an enzyme family called protein kinase C, or PKC. In 1982, Nishizuka and a researcher from France named Monique Castagna discovered that phorbol esters, compounds derived from a natural substance in plants and known to cause skin cancer in mice, target PKC to exert their malignant tendencies.

There was more. As investigators began peering inside tumor cells, again and again they found excess amounts of EGFR, PDGFR, and PKC. The phenomenon came to be known as "overexpression"; genes expressed proteins, and so an excess amount of protein was considered an overexpression. The presence of abnormally high quantities of these kinases strengthened the theory that they were not just linked to cancer but were somehow responsible for it. It was like finding the leaky faucet responsible for a persistent dripping noise. EGFR was found in lung and brain cancers. Excess amounts of PDGFR had been found in several solid tumors and some rare blood cancers.

Then, of course, Matter and Lydon were also seeing the reports on the connection between the Philadelphia chromosome and the Bcr/Abl tyrosine kinase, the fusion protein that resulted when the genes *bcr* and *abl* were brought next to each other in the translocation. Nearly all people with CML had this genetic mutation that created the out-of-control kinase that led to leukemia. Among the growing body of evidence linking kinases to cancer, this one was the best established at the time. People weren't saying that the so-called deregulated kinase caused CML yet, but many were thinking it. Like Druker, Matter and Lydon understood that kinases were being revealed as a driving mechanism of cancer.

At Ciba-Geigy, Matter wanted to start a kinase inhibitor program, a research project dedicated solely to creating compounds "to be used as pharmacological means to get rid of cancer via dysregulated kinases," recalls Matter. Like Lydon, who had stayed at Schering after he left, Matter was convinced that kinases were the perfect drug target.

Matter would need help, though, and who better than Lydon? He had accrued years of knowledge about kinases ever since he'd befriended the postdocs in Phil Cohen's lab, and he was inspired about the possibility of exploiting them to thwart cancer.

In the summer of 1984, right around the time he was reading Tony Hunter's *Scientific American* article "The Proteins of Oncogenes," Lydon received a call from Matter offering him a job with his new kinase inhibitor program at Ciba-Geigy. Lydon wasted no time in making his way to Basel. "He was very courageous," says Matter. "I didn't have a lab, I didn't have technicians. I had nothing."

Matter presented a proposal for a kinase drug program to his supervisors at Ciba-Geigy. The company agreed to fund the project, though begrudgingly. Just a year earlier, Ciba-Geigy had sworn off cancer therapeutics and was refusing to pour any money down what it considered to be an expensive and fruitless drain. But Matter's boss, who was also a good friend, had assured Matter that he could build a cancer portfolio when he came to work there. The best way to do that, Matter was advised, was to keep the program very, very small. The boss also made sure Matter included multiple therapeutic categories in his research department. Alongside the anti-kinase drugs, Matter worked on aromatase inhibitors and phosphonates, two important classes of drugs today. The kinase program was hidden in plain sight, attracting little attention for the moment.

The tectonic plates of cancer care were creaking into motion. The proposal approved by Ciba-Geigy outlined one of the first efforts to design a drug against a specific, well-known target, a sharp contrast to the shot-in-the-dark, trial-and-error approach that characterized the bulk of cancer treatment history. Defiant against the incremental improvements that each new chemotherapy drug had brought, Matter's vision was to make a truly meaningful leap in cancer care by focusing on the kinase. "Today we know [kinases] are so important in the regulation of the growth of . . . cells. It's common knowledge; you learn it in the introduction to cancer biology," said Jürg Zimmermann, a chemist who would soon join Matter's team. "But at that time, there were only a few pioneers who really thought that drug discovery should look into the effect kinases have on the growth of cancer cells." Matter was one of those pioneers.

The approach was called "rational drug design," and the promise of it was huge. As scientists dived into the molecular biology of cancer and began surfacing with concrete results, the notion of fashioning drugs against specific cellular targets materialized as a natural next step. Targeting the root cause of cancer could make for a far more effective drug than chemotherapy. Most chemotherapy drugs carpet-bombed the body in the hope of hitting the cancer cells. If they were aimed toward anything, it was the fastest growing cells in the body, which is why they often led to hair loss, the least debilitating but perhaps most iconic side effect. If individual targets could be identified for each type of cancer, then care could be shaped personally around each patient, side effects would diminish, and the ultimate outcome of treatment—a longer life without disease—would vastly improve.

The moniker "rational drug design" stemmed from the sequence of events by which the medication was created. The end point—say, a haywire kinase—was already known, and the compound was being designed against that predefined result. The approach was the complete reverse of chemotherapy, in which the rationale was far more hypothetical, and it was possible only as a result of accruing evidence that cancer was caused by genetic abnormalities that led to changes in the cellular products of the affected genes. As oncogenes emerged, so did the idea that drugs could be designed specifically against their cancer-causing products in the cell.

The theory behind kinase inhibition was straightforward. With rational drug design, it was the fit that mattered, creating a medication that was perfectly shaped for the surface of its target. Like all proteins, kinases are three-dimensional structures with bumpy surfaces. The ridges and dips on the surface provide cleavage sites, areas for the compound to grab onto, like a mountain climber finding a jutting rock for his next step. The bumpier the surface, the easier it would be to design a drug that would adhere to its target.

No technology was available to see an actual kinase, but its structure could be deduced based on its chemical makeup. All the team had to do was make a compound that would fit onto the kinase, preventing it from binding ATP. Like a gloved hand fitting perfectly over a mouth to block the next breath, the drug would stop the runaway kinase in its tracks, thereby halting cancer progression. Kill the kinase and you kill the cancer.

The approach had its skeptics. The idea of creating a molecule that blocked only one kinase at a time seemed ludicrous to many researchers around the world. If too many kinases were targeted at once, a person could easily die. This kind of selectivity in a drug, the hallmark of targeted therapy, had never even been attempted, so there was no foundation on which to build what seemed like a very faulty tower. And as logical as rational drug design sounded, it was still theoretical. No such drugs had yet been created.

Matter and Lydon were undaunted. The rationale was solid. Plus the field was still relatively simple. In the mid-1980s, when they were getting the project under way, about twelve protein kinases had been found. Creating a compound to selectively block just one of those was sure to be difficult, but it didn't seem impossible. By 1987, about 65 genes were known to encode kinases. Today, 500 kinases have been identified. If Lydon and Matter had known there would turn out to be so many kinases—making the idea of selective inhibition seem preposterous—they might never have started.

Because skepticism loomed large inside the company, Matter had been told to risk-balance his portfolio, to make sure that he was not betting the farm on something everyone else thought would fail. "Marketing had told him that with cisplatin, you could make millions, so what the heck [was he] doing?" Zimmermann said. But Matter, who'd already earned a reputation at Ciba for being headstrong and combative, would not be swayed. Cisplatin was a cytotoxic agent, a drug that poisoned cells throughout the body. It and drugs like it had made cancer treatments notorious for their wretched side effects. Matter and his team were trying to break out of the chemotherapy stronghold, not strengthen it. "His stubbornness allowed him to move forward and follow his mission," said Zimmermann. "He wanted to change the way we did drug discovery in oncology."

By late 1984, evidence was mounting that Matter and his team really were onto something. A Japanese researcher named Hiroyoshi Hidaka had found that a group of compounds called isoquinolinesulfonamides could inhibit kinases. One of his molecules was particularly active against PKC, a kinase that prior research had shown to be overexpressed in some cancers, for example. Ultimately, Hidaka's compounds would not turn out to be effective drugs against cancer (though one was eventually approved in Japan for the treatment of other conditions), but those first reports stirred interest. There were still more skeptics than cheerleaders, but the idea was getting some traction.

By late 1985, Matter's group was growing. Peter Traxler, a chemist who'd been researching antibiotics at Ciba-Geigy since 1973, was moved to the project when the company decided to stop its antibiotics program. The biologists worked under Lydon's supervision, and the chemists under Traxler's. The most noteworthy other additions were Elisabeth Buchdunger, who worked with Lydon, and Zimmermann, who worked with Traxler. A chemist named Thomas Mayer joined the group and worked with Zimmermann on creating the first "hits," compounds that showed some anti-kinase activity, and a scientist named Helmut Mett also came on board. Matter gradually inherited other Ciba employees who, like Traxler, were being moved off of other projects. They were a patchwork of researchers who together created a team that was low-key enough to not attract too much attention from the higher-ups at Ciba.

They were also getting help from experts outside of Ciba. By the early 1980s, Basel had become a hive of biomedical research, in part because of the collection of talent housed at the Friedrich Miescher Institute, or FMI. The facility had been created in the 1970s by then-separate pharmaceutical companies Ciba and Geigy, but it operated independently of industry, doing basic research that could then inform the lab work and administrative decisions at the corporations. Not long after arriving in Basel, Lydon found out that Brian Hemmings, a former postdoc from Phil Cohen's lab at Dundee with whom Lydon had mulled over kinases at the local pub, was now a scientist at FMI. Years later, with the scenery slightly changed, they resumed the conversation. "We spent a lot of time over beers discussing [kinases]," said Lydon, "including would Abl be a good target."

16

A MACHINE WITH A VIRUS FOR A MOTOR

Because this rational design approach to drug development was entirely new, there were no methods in place for doing it. So Lydon's first task was to create a method for examining the compounds created by Ciba's chemists. It was well and good to create a molecular structure that one thought would fit onto the kinase inside the cell. But with no way to test whether it would do anything to cells in a laboratory setting, what was the point? The team needed a way to screen compounds for activity against the stated drug target. That target could be any kinase that was implicated in the development of cancer or other serious diseases. This screening method had to work with whatever experimental molecule they threw at it, and it needed to work with all of the kinases that held promise as potential drug targets.

The techniques available at the time, like gel electrophoresis, the tool Owen Witte had used to separate the Gag/Abl protein into its constituent molecules, were arduous and unpleasant. Today, hundreds of experimental compounds can be checked very quickly for activity against numerous targets at once. Any time a new potential target is found inside tumor cells, drug companies can use high-throughput screening to scan their libraries of experimental compounds for hits. A plate with hundreds to thousands of tiny wells in which a drug target and experimental compound are mixed can be tested by a computer that measures properties that are altered when the target is reacting to a drug. For example, a program might measure how reflective a protein is, because reflectivity increases when a protein is bound by a drug. But back when Lydon and Matter were starting out, there was no technology or protocol for testing drug candidates against a specific target because drugs had never been designed this way before. They had to invent a way to do it.

The challenge was producing large amounts of whatever protein they were trying to target. They needed the protein in large amounts, and it had to be active, still functioning as it would inside the body. Only by having the actual kinase available in cell culture could they determine whether one of the chemists' creations stopped it from phosphorylating protein. Those tests had to be done on isolated enzymes before they could be done on actual cells. But active kinase, the thing that supposedly launched the cascade of events that led to cancer, existed only inside cells. The problem was similar to the one Naomi Rosenberg had solved with her transformation system, bringing the inner, molecular world of the cell into the exposed world of the laboratory. Just as Rosenberg had found a way to study mouse cells outside of a mouse, Lydon and Matter needed to find a way to produce a high-quality enzyme outside of a cell that could be used day in, day out for as long as it took to test anti-enzyme activity of their experimental molecules.

And that was when the trouble began. "It was painstakingly slow to develop this active enzyme," Matter said. "Everything went wrong that could go wrong."

Eventually, the group succeeded in producing live, active kinase using *Escherichia coli*, the bacteria better known as a feared food contaminant. The assay was such a breakthrough that Lydon published the method, a rare move for someone in industry, where secrecy is the default setting.

But the breakthrough was a step, not a leap. The problem with the *E. coli* assay was that it only worked with the Abl kinase. The well-defined link between Abl and CML made this kinase an attractive drug target; there was little question about its importance in the development of the cancer. But that didn't necessarily make Abl attractive to a pharmaceutical company. The problem was that CML was a rare disease. With an annual incidence of about 5,000 people per year in the United States and between approximately 70,000 and 140,000 people worldwide (one to two per 100,000 people), CML was relatively insignificant to a large pharmaceutical company. A drug to treat this rare cancer would have an extremely small "market," as the patient population is referred to. The company needed to be sure that whatever medication eventually came out of this program would reach as many people as possible. A drug for a more common cancer would also be far more lucrative.

For those reasons, kinases that had been found in more common cancers held much more interest than Abl. Three other kinases that had also made their appearance in the scientific literature—PKC, PDGFR, and EGFR—were turning up everywhere. PDGFR was expressed in almost every type of cancer. EGFR had also been unearthed in major cancer types, including lung cancer, a disease with an annual incidence of more than a million people. PKC and EGFR had been found in breast cancer. By the mid-1980s, breast cancer diagnoses had climbed to 350 per 100,000 women in the United States, with about 1.5 million diagnoses worldwide. What's more, emerging cardiology research was implicating PDGFR in a common complication following the insertion of balloons to treat blocked arteries. The idea of creating a kinase inhibitor for cancers that struck hundreds of thousands of people per year, and maybe even for heart disease, captivated the imagination of the research team and the business planners.

The team members needed a way to screen potential compounds for activity against those other kinases because that, they believed, was where the real promise—and profits—lay. They wanted to keep Abl in the mix because its connection to CML was still the best defined in kinase research, but the *E. coli* test alone was far too narrow. They needed to test drug candidates against many more kinases. They'd made a butterfly net when what they needed was a fishing trawl. To make any headway in the search for "hits"—that is, compounds that reached their intended target, accomplishing the goal they were after—they would need a far more powerful assay, one that could screen their experimental molecules against PDGFR, EGFR, and PKC. They needed a way to produce those more appealing enzymes outside of the cell, but no one had any idea of how to do that. They needed help.

Seeking out experts in kinases led Lydon and Matter to Chuck Stiles, a pioneer in the study of PDGFR, at Dana-Farber. Stiles in turn introduced them to Tom Roberts, who had found a way to use baculovirus, a rod-shaped DNA virus, to study tyrosine kinases outside of the cell. By inserting the kinase-encoding gene into the baculovirus genome, Roberts induced the virus to produce copious amounts of kinase. Essentially, he'd created a machine with a virus for a motor that churned out kinase, whatever kind of kinase they wanted. It was the ideal system for studying kinase inhibitors.

The Ciba-Geigy team began collaborating with Stiles and Roberts, and the baculovirus system became their fundamental tool to identify hits. The team also consulted with Robert Weinberg, who had, early on, supported the concept of inhibiting signaling pathways with drugs. Now, with method and advisors in place, the team could move forward with the task at hand: finding a molecule that inhibited the kinase. "And that was when we got seriously going," said Matter.

17

PLUCKING THE LOW-HANGING FRUIT

By the time Lydon and Matter found a way to screen new compounds in the late 1980s, Druker had become a competent scientist. "I was less of a burden in the lab," he said. "Now, instead of asking every single stupid question, I was a little bit more on my own."

He knew about the interactions between his mentor, Tom Roberts, and the industry people from Switzerland, though they were happening only in his peripheral view. In 1987, Ciba-Geigy had sent Elisabeth Buchdunger, the cell biologist Matter had hired, to learn about tyrosine kinase signaling from Chuck Stiles. In the meantime, Druker was continuing his work unpacking the signaling pathway triggered by the polyomavirus that caused cells to become cancerous, a stream of events that also included kinases.

It was right around this time that Druker finally had the 4G10 antibody he'd spent two years trying to make. 4G10 would allow him to measure the amount of phosphotyrosine—that is, the amount of tyrosine onto which phosphates had been bound by kinase—in a sample of cells. The amount of phosphotyrosine revealed by the antibody reflected the amount of active kinase. It was like counting the number of sand castles to estimate how many children had been at the beach.

The antibody intrigued the Ciba-Geigy team, and they wanted to know more about it. Could they use it in their search for drug candidates? Where could they get it? An antibody to phosphorylated tyrosine could be a valuable research tool for someone making a tyrosine kinase inhibitor. It was an ideal way to test the power of potential drug candidates. If a candidate was effective, then there would be less phosphorylated protein. Cells could be exposed to an experimental compound, and then 4G10 could be used to measure the amount of phosphotyrosine in those cells. If the inhibitor was working, the amount would be far lower than in unexposed cells. It was the same thinking that had, in part, led Druker to make the antibody in the first place. The Ciba-Geigy team wanted 4G10. All eyes turned to its creator.

In 1988, after Buchdunger's stay at Dana-Farber, it was Lydon's turn. He loved the collaboration with academia. Lydon thrilled at the freedom of academic research to pursue an interest without a profit-oriented goal, to move about the lab without the lingering shadow of company bureaucracy, and to push experiments forward without waiting for approval. He had chosen to work in industry because he wanted to make new medications, but he relished a chance for a holiday. "What you miss in industrial research is the excitement and fast-moving pace of academic research," said Lydon. "Without that, you just get out of touch in industry because it's very isolated. You can't talk about your work [because] it's all proprietary."

He was also eager to learn about 4G10 from Druker, the pleasant postdoc in the Roberts lab who, Lydon noted, had a medical degree, still a rarity in an academic research lab. Druker's qualifications stood out because Lydon knew that if they found a drug candidate, eventually they'd need a clinician to test it out on cells and, if they were lucky, on actual patients. From that vantage point, an MD who understood kinase science was rare and appealing. Knowing the potential value of the antibody, Druker was happy to share it. Industry-academia collaborations were still relatively simple at the time, the process not yet fraught with material transfer agreements—the legal process guiding the movement of substances in and out of the academic laboratory—and other administrative tasks that inevitably slow the work and cause tension. "Of course you're going to supply it to them," Druker, who has earned a small amount of money from 4G10 over several decades, said of his attitude at the time. This antibody, which he'd created for the sole purpose of learning a new skill that might be useful, now brought him face to face with Nick Lydon.

Two kinase enthusiasts, they had plenty to talk about. Their conversations revolved around the intricate details of what it would take to develop a kinase-blocking drug: what the chemists at Ciba needed to do, what qualities the compound had to have, how it could be tested in cancer cells.

During their talks, they kept coming around to the Bcr/Abl tyrosine kinase, the one implicated in CML. At Ciba-Geigy, the chemists were screening compounds, looking for hits against PKC, PDGFR, EGFR, Abl, and other kinases that had been found to be overexpressed in various types of cancer. PKC, PDGFR, and EGFR attracted the most attention. They were the headliners, whereas Abl was more of a sideshow, a curiosity. Abl—and it was only the Abl enzyme being screened, not Bcr/Abl, which existed only in CML cells—was still the least promising in terms of the target population size.

But by 1988, Druker and Lydon kept circling back around to Abl when they talked about kinase inhibitors. They were convinced that the Bcr/Abl tyrosine kinase led directly to CML. The final proof showing that Bcr/Abl *caused* CML, that the link wasn't just a loose association, would not come until 1990. But the reports from Witte, Canaani, Heisterkamp, Grosveld, and Groffen were enough evidence for them that the fusion kinase caused the leukemia. And the more they talked, the more they started to think that Bcr/Abl was their best shot to test the idea of tyrosine kinase inhibition. At this unproven stage of research, the certainty of the kinase/cancer connection seemed far more important than the commonness of the disease. And the concrete connection between Bcr/Abl and CML did not exist with any other kinase. As Lydon and Druker mulled over the science, Abl moved from an opening act to the main event. By 1989, Druker was adamant that the drug development effort should focus on Bcr/Abl. "CML is going to be the first to fall to this kinase inhibitor," Druker told Lydon.

Lydon concurred. Bcr/Abl was "the low-hanging fruit of the oncogene era," he says. He knew that, as a rare disease, CML was less appealing from a marketing perspective. Compared with the number of people who would buy a breast cancer or heart disease drug, the number of people who would buy a CML drug was tiny. But the clarity of the connection meant this kinase could be the perfect target for proving the principle behind the strategy.

Not everyone involved with the kinase program was so convinced of the importance of focusing on Bcr/Abl, least of all the executives at the company. The marketing projections combined with the skepticism that still clung to the entire effort negated the possibility of focusing exclusively on Bcr/Abl. Matter's team was instructed to keep testing molecules against any kinase, paying the most attention to PKC and PDGFR.

18

A DRUG IN SEARCH OF A DISEASE

The chemists at Ciba-Geigy began churning out compound after compound. They weren't drugs yet; that title could be awarded only after the compound was known to effect some change in the body, not just in a culture of enzymes. During this stage of searching for hits, the chemicals were compounds, candidates, or, if they showed hints of anti-kinase activity, agents. And Matter zealously pushed the chemists to come up with more and more of them, a pressure that Jürg Zimmermann welcomed.

Zimmermann had grown up on a small farm in the Swiss Alps with plenty of time to be lost in his thoughts while he minded the cows and goats in the field. "It was a beautiful life," he said. "Then I started to feel bored." At the age of ten, he began showing an interest in science, and his teacher offered him extra lessons during lunch break. The teacher introduced Zimmermann to the lab, showing him how to do experiments, how to make plastic, how to change the colors of things using chemicals. Zimmermann was hooked. "I saw an opportunity to get away from farming stuff," he said, "and it was fascinating, trying to understand what was happening in the world."

In 1974, at age sixteen, he finished school and took a three-year apprenticeship at Ciba-Geigy, where his work focused on coming up with new types of plastic. Though he loved the work, he didn't like having to follow someone else's instructions all the time. "I was already very interested in inventing something, in designing experiments by myself," he recalled.

Zimmermann left the company in 1977 to obtain a degree in chemistry in Zurich. Academia fed his hunger for knowledge. After studying chemical engineering for three years, he studied organic chemistry for four years, followed by another four years obtaining his doctoral degree in Australia and then Canada. "I still think this was the best time of my life," he said. Although he would try to fit in skiing and mountain climbing when he could, his best weekends, he said, were the ones that left him time to read.

At the end of those eleven years, Zimmermann faced the decision of whether to become a professor or join a pharmaceutical company. He'd grown frustrated doing experiments simply as an academic exercise. Molecules that he and his fellow researchers had synthesized would be thrown away without being tested for any potential application, the sole goal being to prove that the structure they'd created was the one they'd intended to create. "I always protested," he recalled. "Wouldn't it be nice to synthesize something that could be [useful]?" After all, that was what he'd loved about chemistry in the first place: You make something. Just as Lydon had gravitated toward the practicality of kinase research, Zimmermann wanted to be in a place where he could put that urge for application to good use. He opted to return to Ciba-Geigy, joining the oncology group.

Alex Matter opened Zimmermann's eyes to the world of cancer outside the lab. He told the chemists of patients' suffering, motivating them to create the new drugs that were so desperately needed. The understanding that Matter brought to Zimmermann, isolated in the lab as he was, were enough to sustain his determination to do what Matter was asking them to do. In fact, it was a dream come true. "Imagine how it is for a young scientist who wants to make something, who wants to achieve something, who wants to make a difference, [to be told] 'That's exactly the reason why I hired you, we want to make a difference,'" said Zimmermann.

• • •

An Israeli scientist named Alexander Levitzki gave the field of anti-kinase drug research its next push forward. Levitzki had shown that staurosporine, a naturally occurring antifungal agent made by bacteria, blocked PKC, a kinase known to be involved in several cancers. Levitzki's 1986 report became a turning point. "It was the discovery that staurosporine [inhibited] PKC that really made the pharmaceutical industry sit up and take notice," wrote Sir Philip Cohen in a 2002 review of kinase drug development.

The only problem was that staurosporine wasn't specific enough. That is, it blocked PKC, but it also blocked other kinases. To treat cancer, a kinase inhibitor would have to block a single kinase and only that kinase. Because so many bodily processes involve kinases, the drug had to target only the dangerous one in order to avoid serious toxicities like liver, kidney, or even heart failure. This feature of the staurosporine work was fuel for the fire of the skeptics who said it couldn't be done. Kinases were too similar to one another, the thinking went, and they were all competing with each other for ATP, the keeper of energy inside each bodily cell, rendering futile any attempt to take aim at one single enzyme. "A myth therefore began to permeate the field that it was 'impossible' to develop protein-kinase inhibitors with the requisite potency and specificity," wrote Cohen. The inspiration spurred by staurosporine quickly began to deflate. Its trajectory followed the same path as so many other early pharmaceutical bloomers: hype in the wake of encouraging results in the lab, followed by shattered hopes and cynicism when the compound's flaws emerged. Under this lingering cloud of nay-saying, the Ciba-Geigy chemists continued creating molecules that they thought might block PKC, EGFR, PDGFR, or Bcr/Abl.

Their strategy was to focus on the exact place on the kinase that bound phosphate on its way from ATP to the protein. The theory behind kinase inhibition was that each kinase had a particular notch or groove that made a tight fit with ATP, and that the shape of these notches varied from kinase to kinase. That variation was what made kinases viable drug targets. Zimmermann believed that this binding site—the exact spot where the kinase bound ATP to capture the phosphate that would be used to switch on another protein—seemed like the most likely location of each kinase's unique fingerprint. That binding site was what allowed the kinase to serve its function on the cell, so it made sense to Zimmermann that this area would be distinct both from the binding site of other kinases and from other areas on the same kinase. There were no actual images of the molecule inside the cell, but Zimmermann knew the chemistry and could make an educated guess about the chemical makeup of that binding site notch. And knowing this molecular structure enabled him and the other chemists to think about what kind of chemical could prevent it from working. If they knew the shape of the gears, they could throw the right wrench into the clockwork.

The first step in creating a possible drug was to design a molecule that might fit into that slot. If that molecule adhered to the binding site on the kinase, then ATP wouldn't be able to attach there, and the kinase would not get its phosphate. Minus the arrival of a phosphate, the next protein would never be switched on to perform its function, and the cascade of signals that led to the development of cancer would not occur. The kinase would never get a chance to set cancer in motion if it couldn't get to the phosphate in the first place.

Once they had an idea of the shape they were going for, the team could think about how to make a molecule that fit the bill. Zimmermann, Traxler, and the other chemists explored all manner of chemicals that were already known to inhibit certain cellular processes. The art of Zimmermann's chemistry lay in figuring out the combination of elements—carbon, oxygen, hydrogen, nitrogen, and less common ones like fluorine—that would block the site at which the kinase bound to ATP, and how to make that compound. Starting with substances that had already shown anti-kinase tendencies and sketching out their ideal molecular structures on paper, Zimmermann and the other chemists began introducing other atoms into the mix.

Several molecules had already shown some anti-kinase activity (albeit with the coarse profiling tools available at the time). There was the isoquinolinesulfonamide from Hidaka's work. A group from Japan found that a molecule they named erbstatin inhibited EGFR. A group in Israel, led by Yosef Graziani, showed that quercetin, part of a naturally occurring group of chemicals known as flavones, also affected kinase activity inside some tumor cells. The same had been seen with some isoflavones, which also occurred naturally. And there was staurosporine, the antifungal agent that Levitzki had been exploring as an inhibitor of PKC. The fatal flaw of that compound had been its lack of specificity. Could it be adjusted in a way that led it to one kinase, and one kinase only, inside the cell?

The next task was to actually make the chemical, which might require fifteen steps. If the chemists wanted to engineer an interaction between carbon and fluorine, they could follow the rules of chemistry to make that happen. Mix chemical A with chemical B to make chemical C. Mix chemical C with some commercially available reagent to create chemical D. They were creating recipes for new compounds. "Then hopefully at the end you have synthesized a molecule that resembles the one you had drawn on your piece of paper," Zimmermann said.

Zimmermann and Mayer had been asked to focus specifically on PKC, while others on the team homed in on other targets. "People said it's hopeless, forget it," Zimmermann recalled. In the cafeteria during lunch, his colleagues would goad him on, incredulous that he would spend his time on such a task. But Zimmermann didn't see things that way. He'd been given a task, and so he would give it a try. Along the way, he rarely stopped to wonder whether such specificity was feasible; he had just assumed it was. "I just didn't know that it was extremely difficult," he said.

For Zimmermann, the properties of the elements—the explosive nature of hydrogen, the extraordinary versatility of carbon, the adherence of water—had always been more than titillating knowledge. They were ways to make things happen. The rings of weightless, negatively charged electrons surrounding the core of an atom of one element often had space for electrons of an atom of another element. A molecule—a cohesive mix of atoms with its own unique properties—might contain a hydrogen atom, which another atom with an abundance of electrons would stick to as if with glue. Some molecules dissolved in fat rather than water. Inside an oily environment, two chemicals might join together, though such bonds were usually weak. For Zimmermann, these hidden worlds held endless possibilities.

Combining molecules together in various ways had led to the creation of plastic, of countless medications, of every synthetic substance. The manmade world was made of chemistry. So was the natural world, for that matter. Surely, Zimmermann thought, there was a way to manipulate some molecules into a kinase-inhibiting drug.

The challenge lay in more than creating the perfect shape. To be a good inhibitor, the compound also had to stick to the kinase. Creating that snug fit between a molecule and the enzyme was an achievement, but it was no guarantee that the molecule would stay there.

For his molecule to be a viable candidate for a drug, that bond had to be strong. "The stronger the bond, the less of the substance you have to administer to the patient later on," he explained. A compound that was weak might work only at an impossibly high dose. A compound that adhered well to the kinase—that was potent, in other words—would work at a lower dose.

When the chemists managed to create a compound that was both selective—adhering to one specific kinase only—and potent, they sent it to Elisabeth Buchdunger and the other biologists who were part of the team. The biologists had to test each candidate to see if it was active in cancer cells. A molecule might fit into the ATP binding site of one kinase and not any others, and it might stick like cement, but ultimately that meant nothing. The compound had to result in the death of cancer cells; otherwise it was worthless. To become a drug candidate, a compound had to be selective, potent, and active. For the first few years of the program, molecules were handed back and forth, from the chemists to the biologists and back again, each round a new attempt at achieving all three qualities—each time, a moment of starting over, adjusting, and trying again.

Not surprisingly, Buchdunger's tests usually rejected the candidate. A successful compound had to break through the membrane of a cell, make its way through the cytoplasm inside, find the target kinase, attach to the kinase and stay there long enough to kill the cell. Month after month, Zimmermann or the other chemists sent potent and selective compounds to Buchdunger only to discover that they weren't active. Adjusting the molecule so that it might kill the cell often led to a loss in potency or selectivity, and they would have to start all over again.

Among the already existing chemicals the chemists looked to was one called 2-phenylaminopyrimidine, a compound known for its anti-inflammatory effects. When Zimmermann and Mayer tested it against PKC, one of the three kinases in which the company was most interested, the enzyme was blocked. But the effect was too weak; a drug version of the 2-phenylaminopyrimidine would have required too large a dose—grams, rather than the usual milligrams—to be practicable. The body cannot handle such high quantities of a powerful chemical; and even if that were not the case, the administration of a ridiculously large pill or lengthy infusion would be impossible, even parceled out over a day. But having seen that the chemical did something, the team deemed this the "lead" compound, the one it should focus on.

Lead in hand, the chemists had to find a way to improve it. Adding a molecule called 3'-pyridyl to the original chemical scaffolding heightened its activity. With the introduction of a single six-sided molecule, suddenly Buchdunger could see that the compound was blocking PKC with much greater efficiency. Next the chemists added a benzamide group, a molecule created by exposing a version of benzoyl (a component of the acne-fighting combination benzoyl peroxide) to ammonia, and sent the molecule back to the biologists. This time, the molecule was even more active against multiple tyrosine kinases, including PDGFR and Abl.

Then the experiment took a surprising turn. The chemists introduced a methyl group, a combination of carbon and hydrogen, to the growing molecule. In fact, it was just a portion of a methyl group, a so-called flag methyl, which the chemists stuck onto an open slot in the middle of the original backbone. In the diagram of its chemical structure, the molecule now had a single wayward line jutting out from one corner of one hexagon. That single scrap of chemical transformed the entire molecule.

When Buchdunger screened the compound again, she noticed that it no longer inhibited PKC. The flag methyl had eliminated that effect. Before the flag methyl was attached, there were a few different ways in which the atom could form into a single cohesive molecule—several conformations, in chemistry terms. The flag methyl group forced the compound into one immovable arrangement, like a host finalizing the seating for a dinner party. And in that inflexible setup, the molecule couldn't bind to PKC anymore.

The compound was still active—incredibly so. But now that strong activity was directed against Abl. The candidate had been their lead because of its activity against PKC. Now it had become the potent tyrosine kinase inhibitor the team had dreamed of creating—against Abl, the CML-driving kinase. "Just by doing a minor change in the structure, the activity of the compound changed from a PKC to an Abl inhibitor," said Traxler. The compound also blocked PDGFR, though with less strength.

The project's requirements were satisfied: The experimental molecule was selective for a particular kinase, it was a potent inhibitor of that specific kinase, and it was active against cells. Yet the team wasn't quite sure what to make of this new chemical. Here they had designed a compound that exactly matched the one they'd envisioned. But it was for the wrong kinase. Now what? "We've got a kinase, we've got an inhibitor," Brian Hemmings recalled saying when discussing the results with the team. "All we need now is a disease."

The timing was extraordinary. It was 1990, and the final proof that the Philadelphia chromosome, and it alone, caused CML had been made. Druker, Lydon, Matter, and others had already accepted this notion after the evidence had amassed several years earlier. But in strict scientific terms, the mutation had not been proved to be the single trigger for CML.

A man named George Daley, yet another member of the Baltimore lab, had finally accomplished this remaining feat. Daley took one group of mice and filled their marrow with the mutant *bcr/abl* gene present in the Philadelphia chromosome. Next, he destroyed the bone marrow of a second group of mice with radiation. He injected the second group of mice with the marrow from the first group, and the second group of mice developed CML. The experiment established the mutant chromosome, and therefore its protein product, Bcr/Abl, as the sole cause of CML.

The proof bolstered Lydon's belief that the kinase program at Ciba-Geigy should make Abl its top priority. Despite his and Druker's conviction that Bcr/Abl was the best target for proving the principle of kinase inhibition, the program had maintained a general focus on all cancer-associated kinases. The enzymes that were considered to be more desirable targets were those associated with diseases far more common than CML. Now the scientific literature and the company's efforts had converged, with Bcr/Abl as their meeting point.

"That was exactly the time when Nick [Lydon] came and said, 'Look, Bcr/Abl kinase is a very hot item,'" Traxler recalls. "So we changed from a PKC project to a Bcr/Abl project." Lydon knew that CML was a rare disease and that its rareness made an anti-Abl drug less desirable to a pharmaceutical company. But he also knew that inhibiting Abl was the company's best shot at proving the principle of kinase inhibition as a treatment for cancer. Because Bcr/Abl was solely responsible for CML, this cancer provided an ideal way to test out the idea. If a patient's CML stopped progressing as a result of taking an Abl inhibitor, then the company could be assured that the drug was responsible for that change because there were no other cancer-causing factors at play. CML was the perfect testing ground for kinase inhibition, for rational drug design, for treating cancer as a genetic disease.

Now Lydon had the perfect disease for proving the principle of kinase inhibition, and the perfect molecule. The chemists made one last addition to the anti-Abl molecule they'd created. A molecule called N-methylpiperazine improved the compound's water solubility, turning it into a medicine that could be taken by mouth. They sent it to Buchdunger to test for activity.

It worked. At last, about six years after Ciba-Geigy had greenlighted Matter's proposal for a kinase drug development program, Buchdunger was able to report that the compound was potent, selective, and cellularly active. The final molecular formula was $C_{29}H_{31}N_7O \bullet CH_4SO_3$. Described another way, its designation was 4-[(4-methyl-1-piperazinyl)methyl]-N-[4-methyl-3-[[4-(3-pyridinyl)-2-pyrimidinyl]amino]-phenyl]benzamide methanesulfonate. It was a white to off-white to brownish powder with a molecular mass of 589.7; it also carried the weight of forty-three years of science history. The compound was named CGP-57148B. "To me," said Buchdunger, "it was already quite a little bit of a miracle."

19

TWO ENDINGS

In the late 1980s, as the chemistry team in the Ciba-Geigy oncology research department was piecing together molecules like so many Legos, Brian Druker was still plodding away with his polyomavirus project. The work had gone from an interesting way to learn about molecular biology and oncogenes to an increasingly dull project with no end in sight. He'd had a handful of publications come out of his work, but nothing all that gripping. "I was five years in the lab," said Druker. "I didn't have a lot to show for it." By 1990, just when he was getting married and settling down in a house in the Boston suburbs, he was starting to wonder what he was doing with his life.

He'd kept up with his weekly clinic at Nashoba Community Hospital, and he knew that ultimately he wanted to help patients. That had been the goal of his immersion in basic science all along, however unarticulated it had been. By 1990, as he took stock of his talents and interests, he saw his growing expertise in kinase biology and his expertise in cancer care finally coming together. "Why don't I work on a human disease caused by kinase?" he asked himself.

George Daley had just nailed down the final proof about the cause of CML. Because Druker and Lydon had struck up a friendship founded on their common interest in kinase inhibitors, and because he had an inkling that he might want to do research with such a compound, Druker had been following the Ciba-Geigy team's progress. He knew that Lydon had included Abl as one of the kinases against which new molecules were screened. Now, it was Druker's turn to make a bold move. After all the research, all the patients, all the talks with Nick Lydon, he could no longer stand to be on the sidelines. He wanted to be part of bringing this new kind of cancer treatment into the world. He wanted to make a tyrosine kinase inhibitor to treat CML.

He paid a visit to Jim Griffin, a researcher whose lab was one floor below Tom Roberts's lab, where Druker was working. Griffin was a myeloid biologist and knew the ins and outs of bone marrow and the cancers that grew there. The two of them had already worked together uncovering kinases involved in bone marrow biology. Druker asked Griffin if he would collaborate with him on studying the connection between CML and the signaling pathway triggered by the Bcr/Abl kinase. Griffin agreed.

Until 1990, Druker's seizing of opportunities was of a more passive variety. If a door was open, he'd walk through, but he wouldn't force one ajar. For him, the decision to turn his attention exclusively to CML was different. "That was really the first time in my whole life that I had made a proactive decision," he said. He collected cell samples from CML patients. He looked for phosphorylated proteins in those cells. He charted the signaling pathways activated in CML. "That's what I want to do," he realized. "I'll figure out later where it all leads." Having never worked with an experimental drug before, he had just a vague idea of how to start. He knew the Ciba-Geigy team was on its way to generating some compounds that might block the Bcr/Abl kinase. Perhaps he could test them on cells from actual CML patients.

Just a few months after making this declaration to himself, his plans were squashed. Dana-Farber made an agreement with Sandoz, another pharmaceutical behemoth whose headquarters were on the opposite side of the Rhine from Ciba-Geigy. The companies were rivals in the development of drugs for all manner of illnesses. The agreement gave Sandoz exclusive access to the work going on at several Dana-Farber labs, including the one belonging to Tom Roberts, where Druker was still working.

The arrangement was an increasingly common one between academia and industry. Within the universities, grant-funded research was unraveling the intricacies of cancer, heart disease, reproduction, neurological disorders, psychiatric disorders, sleep, allergies, and on and on. But there would always be a limit on how far an academic researcher could take the work because universities didn't have the resources to, say, turn a new finding into a drug development program. Pharmaceutical companies were designed for just that purpose, but often lacked the raw material of original research that could be refined into a new medication. So contractual agreements sprouted up, giving industry access to the latest promising discoveries and giving universities another way to earn money. But the arrangement only worked if the relevant labs were forbidden from conferring with a company's competitors. And though such agreements were on the rise, the offer from Sandoz to Dana Farber was enormous at the time: $10 million per year for ten years. Overnight, Druker was cut off from communications with Lydon, Matter, and Buchdunger. "We could no longer pursue that relationship," said Druker.

Druker and Griffin decided to write a grant application to Sandoz, outlining how they would test a drug that inhibited the Bcr/Abl kinase. They showed, step by step, how they would take an anti-kinase drug that had proved worthy in the lab and investigate its effects on people. But Sandoz wasn't interested in kinase inhibitors. They would give Druker and Griffin some money to develop tools to study them just in case the company ever took an interest. The company was lukewarm at best about the potential for these drugs to work and turn a profit. Druker and Griffin took whatever money the company was willing to give them to explore how, exactly, they would study a kinase inhibitor, were they to ever have one available to test. For Druker, the early 1990s were dedicated to further parsing of the signaling pathway that flowed downstream from the mutant Bcr/Abl tyrosine kinase. But because of Dana-Farber's contract with Sandoz, accessing the inhibitors being created at Ciba-Geigy was out of the question.

It was Druker's personal life that came to inform his next steps. In 1992, he and his wife divorced. After two years of marriage, he'd been forced to face the truth that his work mattered more to him than his relationship. "I was not the best husband," said Druker. "I got married because I thought it was the thing to do, and [she] was a lovely young woman. But when push came to shove, it was about my work, and so I wasn't very present."

As he and his ex-wife headed in separate directions, he gradually rediscovered his love of the outdoors, something he missed from his San Diego days. He remained in Newton, a few miles outside of Boston, and, following the lead of his athletic roommate, another doctor, he began biking to Dana-Farber each day. The exercise quickly became a form of stress relief, a way to rid himself of the mental buildup of weeks of painstaking and often unsuccessful experiments. Soon biking became integral to surviving the lab, the frustrations, and the continual grief over his patients.

His weekly clinic work continued, spanning a total of seven years. He treated people with all different types of cancer, with all the regimens still focused around chemotherapy, turning frequently to the towering experts at Dana-Farber, essentially the birthplace of modern cancer care, for information on prostate cancer, lung cancer, lymphoma. For the moment, the treatments had stagnated. The breast cancer drug Herceptin, directed against a genetic mutation present in some women, was not yet in clinical trials. Tamoxifen, which blocked estrogen and is today considered the first targeted therapy, was in widespread use. Estrogen, a hormone associated with female sexual traits, had long been associated with a certain variety of breast cancer. Thwarting the excess production of the hormone in patients with that type of breast cancer proved to be an effective treatment. But tamoxifen, though aimed against a specific chemical in the body, didn't tackle cancer at its roots. Estrogen is a hormone, not a genetic mutation. A drug inhibiting the production of estrogen could stop or diminish breast cancer growth by removing this chemical on which it depended, but the underlying cause—whatever it was that triggered the excess production in the first place—was left untouched. It was like stopping a car by removing the gas pedal instead of the motor. (The drug, though highly beneficial for many patients, also has some severe side effects, including a heightened risk of uterine and endometrial cancer.) In the eyes of doctors, patients, drug makers, and the public at large, cancer had not yet become a genetic disease. Medicine had not yet become personal to the degree of examining an individual's DNA and tailoring treatment accordingly. The principle had not yet been proved.

20

GETTING OUT OF BOSTON

By 1993, six years after arriving at Dana-Farber, Druker was beginning to feel restless and worried. The link between Bcr/Abl and CML was now firmly established and widely known. Around the world, new kinases were continually being discovered. Investigations into the molecular biology of cancer churned out an increasing number of connections to mutant versions of these enzymes. New oncogenes routinely appeared in the literature. One day, Druker would have an idea about how to target a kinase with a drug, and the next day, he'd read about that same idea in the scientific literature. He began to feel trapped by his circumstances. "If I have a good idea I should be able to execute it and be the one that publishes it," he thought. "But I can't do that if I'm the only person doing the work." Griffin had proved to be an excellent collaborator, but he didn't have his mind set on creating a CML drug as fiercely as Druker did.

It was time to start his own lab where he could steer the research and focus exclusively on the development of a kinase-blocking drug for CML. And there was only one way to do that: He had to ask Dana-Farber for an assistant professor position, some funding, and some space. "And that didn't go very well," said Druker.

He made his application to the chief medical officer at Dana-Farber, a man named David Livingston, who like Druker was both a doctor and scientist. He wasn't the top leader at Dana-Farber, but he was the gateway to all faculty appointments. Livingston declined Druker's request for his own lab. "He didn't believe in either me or my work," said Druker. "He didn't think I had what it took to run my own lab."

Livingston presented another option to Druker. Dana-Farber was opening a new molecular diagnostics lab, and he offered Druker the option of running the lab. As scientists and doctors learned of the various mutations cropping up in tumor sequencing, the idea of knowing a patient's mutation status assumed importance. No one knew which abnormalities might be relevant to the progress of a cancer, but including their presence in a cancer patient's medical records seemed a logical next step. Plus, inherited traits associated with cancer were increasingly coming to light (as in the case of Li-Fraumeni syndrome, a hereditary condition that predisposes people to cancer), so the notion of testing families for these genetic conditions was gaining ground. This lab was one of Dana-Farber's first forays in that direction. Livingston thought that running the new operation would take up about half of Druker's time. The other half could be spent pursuing his chosen research.

Druker was intrigued. If half his time was still his own, he reasoned, that could work. He could get somewhere with his kinase research. So he made a list of what he thought the diagnostics lab would need—the technology and other resources that were making this new discipline possible. "I never heard back," said Druker. Apparently, Livingston was offended because Druker had asked for more than the company was willing to give. Poking that hornet's nest made Druker realize that it would have been a fruitless pursuit. "I wasn't going to do something that they're not going to put any money into," he says. "Why would I spend my time and energy on something that was set up to fail?" But, having offended a man so near the top, Druker needed to face the fact that he was not going to be offered a job at Dana-Farber. He had to make a choice: stay in the lab with Tom Roberts or leave one of the world's most respected cancer research institutions and strike out on his own. He knew it had to be the latter.

While Druker was negotiating with Dana-Farber about the diagnostics lab, he'd also applied for a job at Beth Israel Hospital in Boston.

They were setting up a special unit for studying signal transduction, the sequential cascades of signals that flow from one protein to the next within the cell. It was the exact work to which Druker had dedicated the last six years of his life. Druker was offered a position, just as he was waiting to hear back from Dana-Farber about the diagnostics lab. When Druker pressed to get something in writing, the chair of medicine who'd made the offer insisted it wasn't necessary, that the position was his if he wanted it. Druker waited to hear back from Dana-Farber, still hoping he could stay there. When he finally called to accept the offer at Beth Israel, he was told the position was gone. The hospital had already hired someone else.

"So at that point, I've been kicked in the stomach twice," says Druker. "Do I just disappear, or do I get up and decide what I really want to do and make something of this?" Quickly, 1993 was turning into a low point for Druker. His marriage had failed, and the city he'd made his home seemed to have no space for him as a scientist. As far as he could see, there was only one possibility. "I need to find a position," he decided. "I need to get out of Boston."

Opening a private oncology practice was an option, but it was never something Druker wanted. He didn't want to give people chemotherapy. His dream was to fulfill his vision of bringing better treatments to patients. He wanted to make good on the promise he'd written in letters to the families of his patients who'd died. "I was going to go into the lab, and I wasn't going to come out until I had something that was better than what we had to offer," Druker had told them.

Now, that vision had taken on a very specific form: creating a drug for CML that targeted Bcr/Abl. "That was my goal. I knew that was what I wanted to do. If I were going to make a difference, if I were going to live up to the promise I had made my patients, that's what I needed to do," he said. Suddenly, it didn't matter how illustrious the institution was; all that mattered was that he work toward this singular focus. "And it all became really clear."

Colleagues asked Druker why he was taking such a hard road. Surely, they asserted, he would burn out from all the effort, the constant fight for funding, the endless struggle to create a drug that still seemed impossible. But Druker knew that it was the seemingly easy road that would burn him out the fastest. He knew the endless conversations informing patients that there was nothing more he could do would leave him spent. During his years at Nashoba, he was having that talk once every week or so, telling hopeful sufferers that the current drugs were no longer working and that it was time for him to focus on making them as comfortable as possible during their final days. "If I were in practice, that would be [happening] once a day," he says. "There was no bike ride I could take that could get me through that."

So he made a list. He wrote down all the academic institutions with small but growing oncology programs. He figured that a nascent program, rather than one that was fully established, would allow him to have a hand in its development and would give him the freedom to pursue his rogue dream of kinase inhibition. He also made a list of places where he'd like to live. Now that he was leaving, Druker had to admit he'd never felt at home in Boston. "I always felt like an outsider there," he said. Letting go of the fixation on prestige so ingrained within academia made space for other considerations. He began to think about the outdoors. He missed the lifestyle he'd had in San Diego during his early medical studies, and he realized how important exercise had become to handling the stress of his career. He needed to be in a place where he could run and bike year round.

He looked into hospitals in New York, at the renowned Cold Spring Harbor laboratory, at the University of Iowa. Everything was interesting, but nothing was quite right. Then he made a visit to the Oregon Health and Science University, in Portland, to meet a man named Grover Bagby. "I saw a person who was committed to growing a cancer program, who believed that targeted therapies were going to be the future," said Druker, "[and] who I thought I could trust." His first interview happened to fall on a sunny day in January—"I think there's two of them," Druker said dryly—and the mountains and profusion of green were as attractive to him as the job for which he was interviewing. "I just fell in love with this place," he says. So wooed was he that he didn't even think to look at how OHSU was ranked academically. "I guess if I had done that, I would have thought twice," he says. After all, he was coming from one of the world's top academic research institutions. Yet his experience there had clarified that such status was no guarantee of personal success. When Bagby made him an offer soon after his visit, Druker readily, and happily, accepted.

Shortly before he left Dana-Farber, once he knew he was on his way to Portland, Druker called up Nick Lydon. Knowing he was leaving, Druker had no more patience for operating within the confines of the institutional agreement with Sandoz. He wanted to talk to Lydon about kinase inhibitors. He told Lydon his plans and asked if he had any inhibitors for the Bcr/Abl kinase. "As a matter of fact, we do," Lydon told him.

He excitedly informed Druker that the chemists at Ciba-Geigy now had several candidates for kinase inhibition, including one with strong activity against Abl. Since Bcr/Abl existed only inside CML cells, the experimental molecule had been screened only against Abl, the enzyme naturally occurring inside normal cells, and not against the mutant fusion protein that resulted from the Philadelphia chromosome. They had to wait for someone like Druker to test the drug in actual CML cells to know whether it inhibited Bcr/Abl. That work could have been done by anyone, but Lydon had been hoping it would be Druker.

"Would you like to test them?" Lydon asked.

"I can't test them until I move," Druker replied.

"Okay," Lydon said. "Call me when you get to Oregon."

By July 1993, Druker had moved and was setting up his lab. He rented an apartment in Lair Hill, in the southwestern part of the city. He walked to work each morning along a winding road bordered by towering moss-covered trees, their branches a braid of neon green, a view interrupted only by the sight of bikers and runners making their way uphill. Woodpeckers tapped a beat into the cool air of the Pacific Northwest as Druker strode to the top of Marquam Hill. There, the cluster of buildings that housed OHSU boasted views of Mount Hood and Mount St. Helens, their snow-covered sides glistening in shifting shades of pink and yellow on sunnier days.

By August, just a few weeks after his arrival in Portland, Druker had received several compounds from Ciba-Geigy and had begun the experiments he'd been dreaming about for several years.

21

KILLING CELLS

In 1993, Ciba-Geigy's lead compound in the kinase program, the one considered to be the most promising to bring to market, was not the one targeted against the Abl kinase. Rather, it was another compound, similar but with key differences that made it active against the PDGFR kinase. Even though Lydon believed that the anti-Abl kinase for CML was the company's best shot at proving the principle behind kinase inhibition, the rarity of CML failed to generate enthusiasm among the higher-ups. Matter, Lydon, and the rest of the development team had been instructed to continue creating molecules to block kinases involved in more common cancers. That put the PDGFR compound—code-named CGP-53716—out in front. CGP-57148B, the development name for the molecule that blocked Abl so potently, was second in line. Lydon sent Druker these two compounds, two other candidates, and a dummy molecule that had no anti-kinase activity. When Druker received the molecules, they were labeled with the company code names, but there was no information about which kinases the different molecules blocked. He was told nothing about CGP-53716, CGP-57148B, or the other active molecules. He also wasn't told which one was the control, the molecule that the Ciba-Geigy team knew did nothing. Like the sugar-pill placebo often used in human drug studies, the control was there to confirm that any changes seen with the experimental molecules were not also seen with a compound known to have no effect. Because he didn't know the identity of each compound, Druker could conduct a blind study, which legitimized the results. Knowing which compound was active against which target would have introduced a potential for bias that could undermine his lab findings.

Before Druker had unpacked in his new home and settled into his new lab, he was testing Ciba-Geigy's compounds. For the next year and a half or so, Druker, along with one of his two postdocs, focused almost exclusively on testing the Ciba-Geigy compounds. Another postdoc studied Bcr/Abl signaling, research that was more grant-friendly than the work he was doing with Ciba-Geigy. Druker was given a couple of graduate-level classes to teach, but his primary goal was exactly what he'd come to Portland to do: help usher in a new drug for CML that could, if successful, prove revolutionary for the treatment of cancer.

He experimented with CML cells derived from people and on white blood cells taken from mice that had been forcibly engineered to express Bcr/Abl. He also had human cells derived from other types of leukemia that did not contain the mutant kinase. Finally, he had samples of bone marrow from CML patients who'd undergone bone marrow transplants and people who had undergone the transplants for reasons other than CML. He laid out a protocol that would enable him to isolate the effects each experimental compound was having, if any. Much of the work followed the approach he and Jim Griffin had presented in their grant application to Sandoz back in 1990. Druker had to test whether any of the experimental compounds killed the cells, and, if that happened, confirm that the cause of death was the compound and nothing else.

As he had learned to do in Tom Roberts's lab, Druker set up a tray of test tubes. In four of them, he mixed a minuscule amount of CGP-57148B (which, unknown to Druker, was the compound directed against Abl) with cells derived from patients with CML. In another group of tubes, he incubated those same cells with a higher amount of the experimental compound. Yet another group held cells only, with no compound present. He arranged a parallel tray for the mouse cells. Each day, for four days, Druker pulled a sample out of each tube and counted the number of cells. If CGP-57148B was working, the number of cancer cells present should decline markedly each day. Blocking the kinase responsible for the cancer should kill the cancer cell.

Each incubation had started with anywhere from 5,000 to 20,000 cells. At the end of those four days, many of the tubes were still rife with viable cells. In the tubes with human CML cells and nothing else, the number of cells was now around 800,000. The mouse cells had similarly continued to proliferate in the test tubes, and the lower dose of the compound had not appeared to thwart the activity of either type of cell.

But in the tubes where the cells derived from human CML had been incubated with CGP-57148B, something astonishing had occurred: the malignant cells were all dead.

In another group of test tubes, Druker incubated human-derived cells containing the *src* gene, instead of the *bcr/abl* fusion gene. This change enabled Druker to test whether the cell-killing activity of the compound was connected to the presence of Bcr/Abl. In those tubes, the cells kept multiplying. CGP-57148B was selective. In the mouse cells, the results were identical.

Druker then made sure to run a blind round of experiments, one in which a second researcher tested the candidates without knowing the results of Druker's experiments. Even though Druker was already blind to the target of each compound supplied to him by Ciba-Geigy, he wanted to make sure he wasn't inadvertently biasing the results. He dissolved the compounds, gave a postdoc four vials labeled A, B, C, and D, and asked him to test them. He knew that if his study was published, another lab would immediately set out to replicate his research. He also knew how easily error could creep in, for example by counting a cell as dead when it might not be. The blinded approach prevented the postdoc from manipulating the data just to please the boss. When the results revealed that the vial containing CGP-57148B undoubtedly killed CML cells, Druker was confident that anyone who got the compound would reproduce their findings.

"Do you want to join us? Brian has the first results," Lydon asked Zimmermann, back at Ciba-Geigy in Basel. It was early 1994, and Lydon was heading to a conference call with Druker to hear the findings of his initial studies, about three months after Druker had gotten the experiments under way. Some of Druker's experiments had been done weeks earlier, but with his postdoc's verification now complete, Druker was finally ready to report all the results.

Zimmermann joined the table, where Lydon, Matter, Buchdunger, and a handful of others waited to hear the news from Portland.

"Thank you for the shipment," Druker began. "The samples have arrived, I have tested them in my assays, and they seem to work."

Zimmermann still remembers the moment vividly. "That was unbelievable," he recalled.

Elisabeth Buchdunger was also amazed by the news. "When we got back the data that clearly showed that the compound was killing only the cells that had the Bcr/Abl kinase and did not affect normal blood cells, this was a really fantastic thing," she said. "It clearly proved the selectivity."

There was one other piece of information Druker had to share with the group in Switzerland. The compound also inhibited another kinase that had been introduced to the screening tests: Kit. This kinase phosphorylated tyrosine, just as Abl did. Druker had added Kit to the panel of kinases against which the molecules were screened, and CGP-57148B was active against this one, too.

The collaborative team—Druker, his postdoc, and the main chemists and biologists at Ciba-Geigy—wrote a paper describing the study that they'd soon submit to a journal. A publication was essential, they all knew, for staking their claim in the kinase territory. A prominent paper was crucial. It would help position them as leaders in the field, it would ensure that they would not be accused of following someone else's work, and it would create buzz about the increasingly realistic possibility of creating a selective kinase inhibitor. For Druker, the report would be the first tangible testament that his long-held hunch about kinases and the vision of improving cancer care that had propelled him forward was now inching into reality.

Considering the striking results, the team of authors felt confident about placing their work in a prominent journal. They sent the manuscript to *Science,* but it was rejected. Undeterred, they submitted it to *Nature,* another top-tier publication. Again, it was rejected.

Druker found himself simultaneously inspired and frustrated, not only from the journal rejections. The findings from this first preclinical study were strong enough to warrant moving the compound to the next phase of development. "It was exciting, and we were on a pretty linear track to clinical trials," says Druker. But everything was taking too long. He had expected the company to respond to his data by initiating further research immediately. He knew there were patients in need of new treatment options, and he knew this compound could potentially save their lives.

He also knew that researchers around the world were getting more interested in kinase inhibitors, and he didn't want to miss the chance to be the first. "Part of it was I didn't want to get beat," Druker acknowledged. He had reason to be worried. In 1995, while the journal article he and the others had prepared was under review, Alex Levitzki, who'd discovered the anti-kinase activity of staurosporine (the antifungal agent that blocked too many kinases at once), published a report about a kinase-inhibiting compound based on that naturally occurring chemical. The researchers with whom Druker was working feared being scooped. "There were patients, there was competition," he said. "I wanted us to get moving."

He was baffled that Ciba-Geigy, upon seeing the results, was not plowing ahead. The company knew that the next step was to prepare for clinical trials by testing the drug on animals. It would need sufficient amounts of toxicology data from animals if it was to ask the FDA's permission to study the drug in humans. That was the routine process for so-called investigational new drugs, one in which the company was well versed.

"Ciba was being odd about all this," Druker said. He knew that Lydon shared the same frustration, and he tried to be patient. He recognized that some progress was being made, but he also knew there was no time to waste. "It just seemed like, can't you just do this right now? Can't we start our clinical trial next month?" Druker remembered thinking.

The company also knew the signs of a promising drug candidate. CGP-57148B had them all—almost. There was one irreconcilable problem: the limited market. There were too few patients with CML to make developing this new drug worthwhile for the company. To make the investment required to move the drug from animal testing through clinical trials and then into distribution and marketing if the drug was approved by the FDA (a costly process in itself), a pharmaceutical company needed to be assured of profits typically in the hundreds of millions. No way would a drug indicated for the rarely occuring CML bring in that level of sales. There was no way around that fact. The potential efficacy of the drug was intriguing, but it wasn't enough to sway the company away from its concern about the bottom line.

Plus, the mission of the company was to provide medical advancements to the people who needed them. If the company was going to invest its resources in cancer, didn't it make sense to focus on the most widespread types with the highest incidence so that the most people possible could benefit from the research?

"There was a lot of resistance in the Ciba-Geigy marketing department that CML was too small an indication [for] trials," said Lydon. His group was being told to go for PDGFR as the first clinical entry point. "[CGP-57148B] was always the stepchild," says Druker. "It was not something that [the] company was pushing fast to get into the clinic."

Fortunately, another insight from Druker's preclinical study added ammunition to the plea for a clinical trial. Early on, when he and others were still just hypothesizing and daydreaming about targeting kinases, Druker had thought that such a drug might ultimately be given by treating the bone marrow. The concept was that the bone marrow of a person with CML would be removed and incubated with the drug. The cancer cells in the marrow would die, and the normal cells could then expand. This cleaned marrow would then be transplanted back into the patient.

For that reason, Ciba-Geigy had sponsored Druker's lab to test out a second approach: mixing the compound right into the marrow as a way to stop the leukemia. This strategy was why he'd obtained marrow samples to test along with the mouse and human cell lines. Druker had tested both samples—those free of CML and thus the Bcr/Abl kinase, and those from CML patients known to have the

Philadelphia chromosome—with the experimental compounds sent by Lydon.

For these tests, the cells were brought to the cytogenetics lab for analysis. A new microscopy technique known as FISH, short for fluorescence *in situ* hybridization, had just become available. By coaxing specific areas of DNA into glowing one color or another when viewed through a fluorescence microscope, FISH allowed geneticists to obtain a highly accurate count of cells with a particular gene sequence, or rearrangement of a gene sequence. With FISH, *bcr* glows green and *abl* glow red. In a normal cell, the red and green dots will be spaced far apart, signifying their location on separate chromosomes. In CML cells, the red and green dots of the rearranged chromosomes are adjacent, brought together through the translocation of material between chromosomes 9 and 22 that defined the Philadelphia chromosome abnormality. When next to each other, the red and green dots often appear yellow, a result of the properties of the fluorescent light being emitted. Looking through a microscope, a researcher could count exactly how many cells in a particular sample were positive for the Philadelphia chromosome by finding those yellow spots. The invention was incredibly well timed for studying CGP-57148B because it enabled Druker to know the exact extent to which CGP-57148B was affecting the cancer cells. If the compound was doing anything, then the count of cells with red and green dots lumped together would be lower.

Without divulging too much information, Druker would carry vials down to the cytogenetics lab in a circular Styrofoam cooler, their plastic pop-top caps concealing an undisclosed solution mixed with thick clusters of bone marrow. There, Helen Lawce, the resident chromosome expert who'd worked briefly with Joe Hin Tjio (the Indonesian scientist who'd provided the first accurate count of human chromosomes), exposed the cells to fluorescing DNA probes and examined them under the microscope. Her white hair gathered in a ponytail flowing down her back, Lawce sat in a darkened room adjusting the focus on the giant fluorescence microscope, the folk music she preferred playing quietly in the background.

In the non-CML samples, the compound did nothing. But when Lawce looked at the samples that had come from CML patients, she

◀ Peter Nowell, MD, *(left)* and David Hungerford, codiscoverers of the Philadelphia chromosome in 1959.

▶ These microscope photographs are from Nowell and Hungerford's first full-length publication on the Philadelphia chromosome. Both figures show human chromosomes halted during cell division using the innovative process developed by Nowell. Figure 1 is from a healthy subject. In figure 2, an arrow points toward the abnormally small chromosome 22 in a cell from a patient with CML. Although staining techniques were rudimentary, Hungerford was able to spot this mutation—the Philadelphia chromosome.

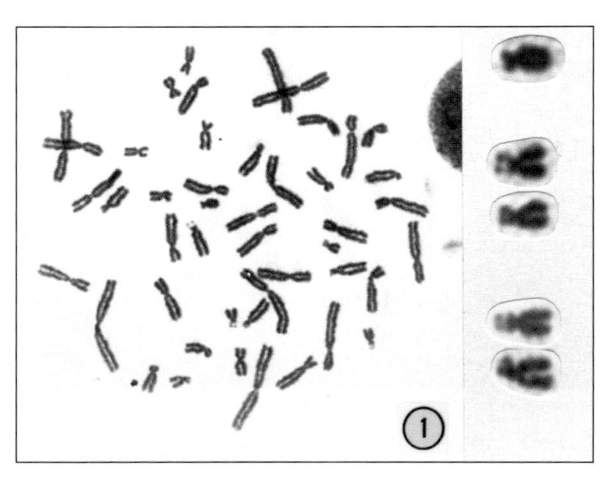

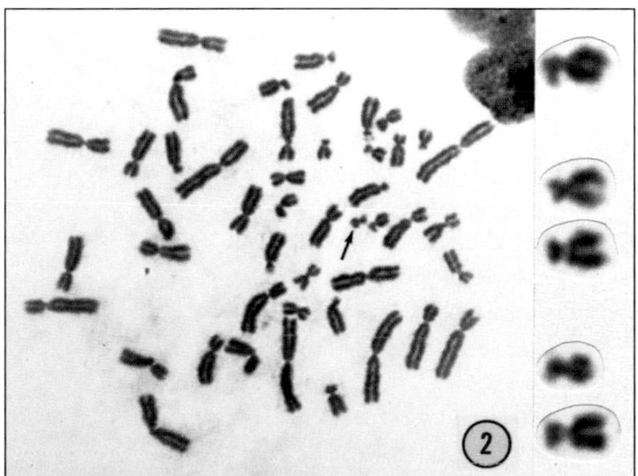

▲ Cell cultures from Naomi Rosenberg's experiments with the Abelson virus. *Left*: Cells unexposed to the virus remain invisible to the naked eye. *Right*: Each black dot is a visible cluster of cells that arose from a single, virus-infected cell.

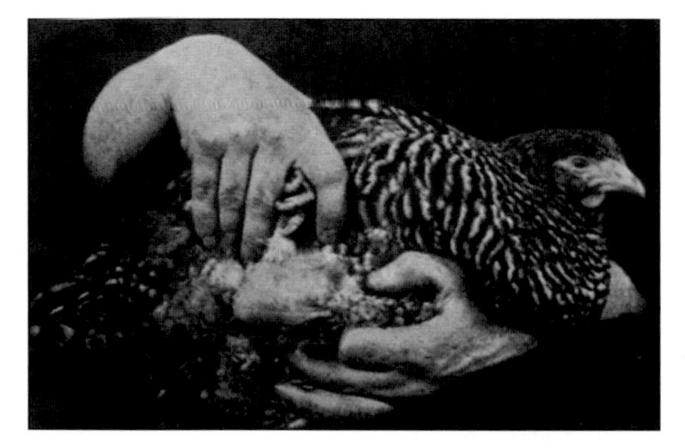

◀ The Barred Plymouth Rock hen brought to Peyton Rous in 1909. The virus isolated from the hen's tumor was instrumental in unraveling how viruses trigger cancer transformation, and ultimately led to the discovery that cancer-causing genes come from healthy genes.

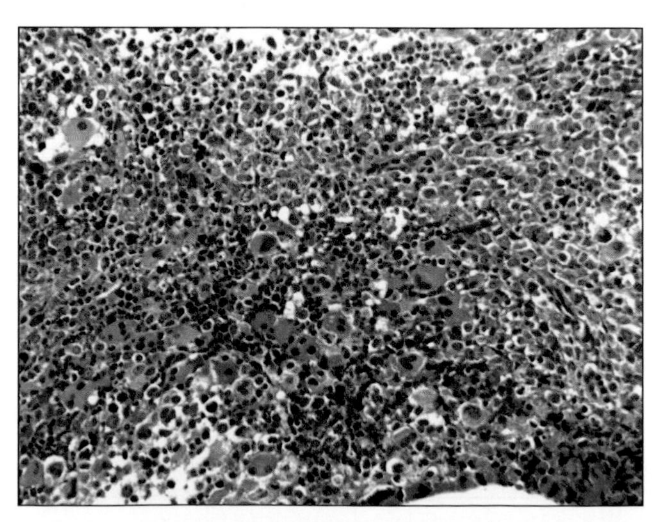

▶ A bone marrow biopsy from a patient with CML. This sample is overcrowded with white blood cells and platelets, making it far more condensed than normal bone marrow. One of the first symptoms of CML is bone pain resulting from this excess proliferation of blood cells inside the marrow.

The *src* probe created by J. Michael Bishop and Harold Varmus that uncovered the cellular origin of oncogenes

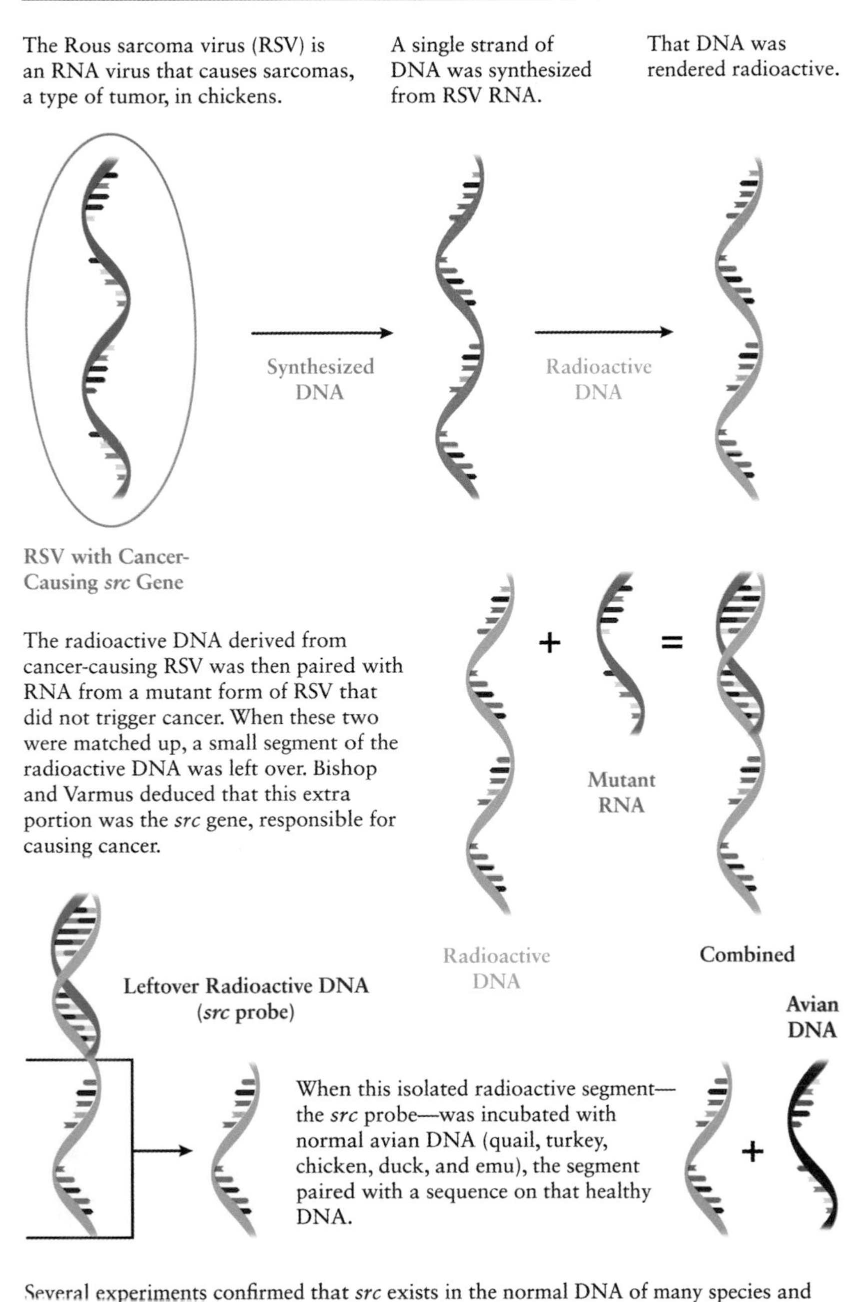

Several experiments confirmed that *src* exists in the normal DNA of many species and that the gene became cancer-causing when it was integrated into the RSV genome during an infection. This research led to the realization that cancer-triggering genes (called oncogenes) are mutated forms of healthy genes.

◀ Janet D. Rowley, MD, who in 1972 discovered that the mutated form of chromosome 22 was part of a translocation, in which a piece of chromosome 22 and a piece of chromosome 9 swap places.

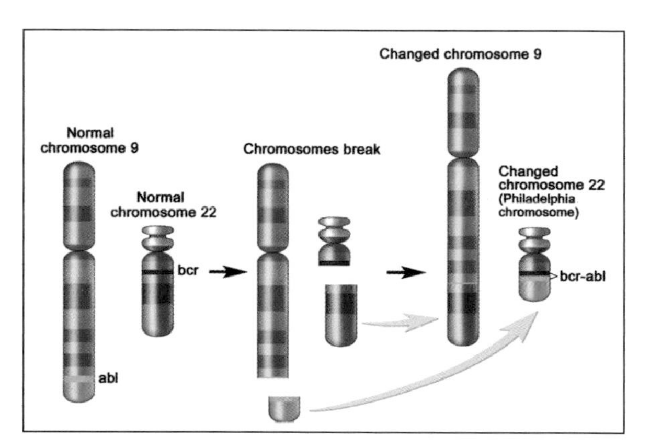

▶ In the Philadelphia chromosome translocation, the *abl* gene, usually on chromosome 9, moves next to *bcr*, on chromosome 22, resulting in a mutant gene, *bcr/abl*. The fusion of these genes, the defining feature of the Philadelphia chromosome, causes chronic myeloid leukemia.

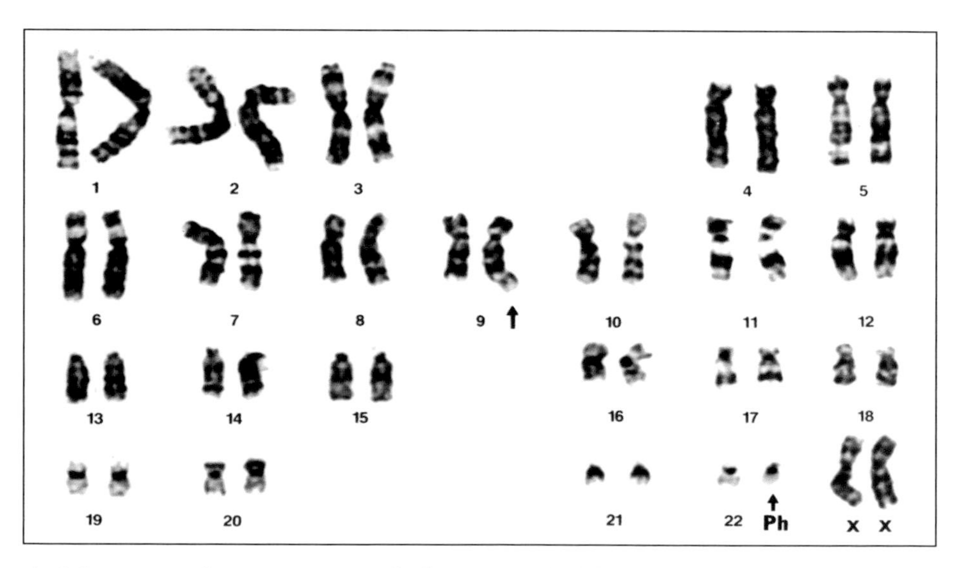

▲ A karyotype from a patient with chronic myeloid leukemia, created with the banding techniques that became available in the 1970s, showing the extended chromosome 9 and truncated chromosome 22 (the Philadelphia chromosome).

From the Philadelphia chromosome to CML, and how kinase inhibitors stop that process

Humans have 23 pairs of chromosomes

Cell Nucleus

In a rare, spontaneous mutation, chromosome 22 swaps a small amount of genetic material with chromosome 9 to become the Philadelphia chromosome.

bcr/abl

This translocation results in a mutant gene, ***bcr/abl.***

Bcr/Abl

The mutant *bcr/abl* gene encodes a mutant protein called **Bcr/Abl.** This protein is a kinase, a type of enzyme that triggers various processes within cells.

Without Kinase Inhibitors

With Kinase Inhibitors

Blocked Phosphate

P

Drug

Normally, the Abl kinase transfers phosphate from ATP (the cell's energy source) to another protein, launching white blood cell production, a process that turns on and off. With the mutant Bcr/Abl kinase, the process never stops.

A T P

P

P

Protein

This haywire activity causes the body to produce too many white blood cells, the hallmark of CML.

Kinase inhibitors block the site where phosphate binds to Bcr/Abl, preventing the haywire kinase activity.

As a result, white blood cell counts are normalized.

White Blood Cells

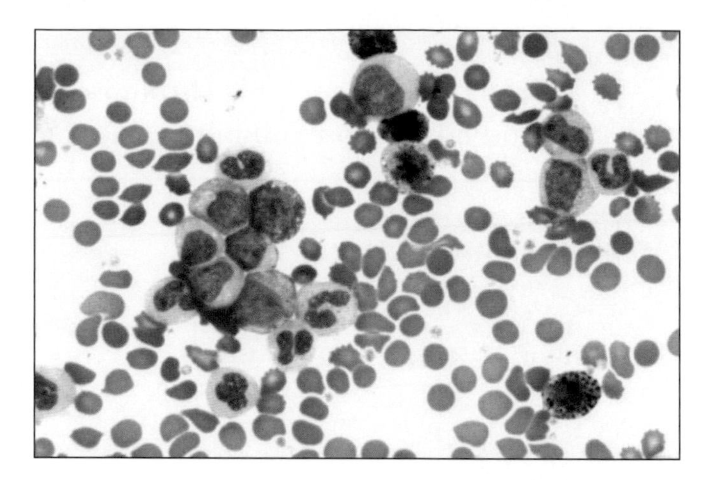

▶ A sample of blood from a patient with CML. The excess abnormal white blood cells that are the hallmark of this disease are stained purple.

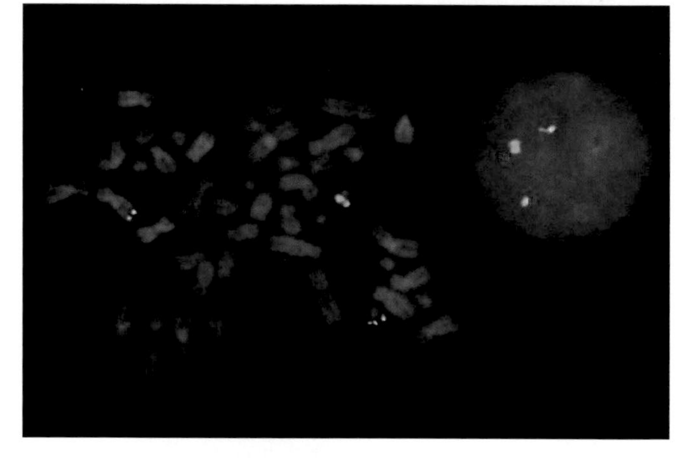

▶ A FISH (fluorescence *in situ* hybridization) image of chromosomes from a patient with CML. The red dots indicate the *abl* gene, and the green dots indicate the *bcr* gene. The yellow dots indicate chromosomes on which *bcr* and *abl* are fused together.

▲ Jürg Zimmermann and Elisabeth Buchdunger of Novartis Pharmaceuticals. Zimmermann and Buchdunger worked together to create the first tyrosine kinase inhibitor. Zimmermann led the effort to synthesize experimental compounds, and Buchdunger tested each new compound for possible anticancer activity. The lead drug candidate that emerged from their work blocked the Bcr/Abl kinase and was named CGP-57148B, then STI-571, and finally Gleevec (imatinib mesylate).

▲ 2000 Warren Alpert Foundation Prize recipients. *From left to right:* Owen Witte, Nicholas Lydon, Brian Druker, Alex Matter, and David Baltimore. The award was given to all five individuals for their research that led to the development of the first tyrosine kinase inhibitor.

◀ Brian Druker and LaDonna Lopossa, the sickest patient to enroll in the STI-571 clinical trial at OHSU. Lopossa had chosen her burial plot shortly before she began treatment in 2000; this photo was taken in 2011.

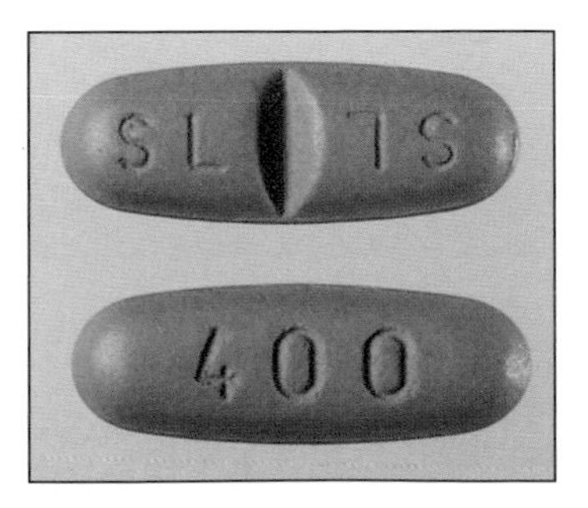

▶ A 400-milligram pill of Gleevec, a typical daily dose for someone with CML.

▲ A 2010 gathering at Fox Chase Cancer Center in Philadelphia to celebrate the 50th anniversary of the discovery of the Philadelphia chromosome (people whose names are in **bold** below appear in the book).

Back row: Felix Mitelman, Alfred Knudson, Joseph Testa, **Peter Nowell, Nicholas Lydon**, William Sellers, **Owen Witte**. *Front row:* **Janet Rowley, Alice Hungerford** (holding a photo of her late husband **David**), **John Goldman, Nora Heisterkamp, Charles Sawyers**, Hope Punnett.

▲ This photograph of Gary Eichner and his son, Tuff, was taken in 2013, about a year after Eichner began treatment for chronic myeloid leukemia.

▼ A 2001 advertisement for Gleevec featuring Suzan McNamara, who led the patient-driven effort to request that Novartis speed up production of the drug in 1999.

was perplexed. She called Dr. Druker to explain her confusion. "I can't tell you how many of these cells are Philadelphia positive because they're all dead," she told him. Druker was elated. Lawce didn't know what to make of his response. "What's that about?" she thought to herself. "I can kill cells with Clorox." Druker was intent on keeping his work under wraps, so for Lawce, his delight remained a mystery. She couldn't understand why it was a good thing that all the cells had died.

But Druker knew their death meant the compound had worked. In this set of experiments, up to 80 percent of the bone marrow cells from CML patients were destroyed by exposure to CGP-57148B. As the remaining cells repopulated, those that did so in the absence of the compound contained the *bcr/abl* gene. Among those remaining exposed to CGP-57148B, fewer than 20 percent contained the mutant gene. Druker also noted that in the marrow sample that came from a CML patient who did not have the Philadelphia chromosome—a rare but possible occurrence—the compound did not impede the growth of the cells.

The marrow data showed the team at Ciba-Geigy two things. First, it revealed that the marrow-cleaning approach, known as purging, wouldn't work because the normal cells would not expand back to a safe quantity. But also, the work proved without a doubt that the compound killed CML cells and spared normal cells. All of the other test tubes provided solid evidence of the compound's potential. But bone marrow from real CML patients was the closest thing to actual patients. If the compound worked with marrow, odds were it would work with people.

"For us, this was a very convincing piece of data that pushed CML to the forefront," said Lydon. He, too, had been frustrated by Ciba-Geigy's hesitation about moving forward to the next phase of development. He was sympathetic, at least to an extent. "It was a commercial organization, and they had to make a drug that could sell," he said. But he also knew that when it came time to pioneer an entirely new type of medicine, the smartest approach would be to test the drug in the area where it was most certain to succeed. "CML was that area," said Lydon. With the marrow data, Lydon, Matter, and the rest of the research team had irrefutable proof that this compound deserved a clinical study. "[This] was the piece of data that we used to convince the organization that CML was the area to test this drug."

Finally, in 1995, the company agreed. That June, Ciba-Geigy held its first meeting to plan the clinical development of CGP-57148B. The meeting was attended by Lydon, Buchdunger, Druker, John Ford—the central medical adviser in the clinical development department at Ciba-Geigy—and a man named Alois Gratwohl, an outside consultant to John Ford, there to advise on the best course of clinical trial action. CGP-57148B was cleared to enter toxicology studies. A phase I clinical trial—the first stage of human testing for any new drug—was tentatively scheduled for November 1996, twelve years after Matter had first started the kinase inhibitor program.

22

GAIN AND LOSS

There were two tasks at hand once CGP-57148B was cleared for development. First, the compound had to be turned into a drug. Second, the drug had to be proved safe. Safety didn't mean the drug would have no side effects, but the company had to be sure it wouldn't kill people.

Moving from the laboratory studies to this early development stage was a huge undertaking. It was one thing to have a molecule that worked inside a cell. It was quite another thing to turn that molecule into a palatable medication that worked inside a living, breathing human, particularly one who was sick with leukemia. The immune system isn't the body's only guardian against foreign substances. The digestive system is also geared to reject anything that has no benefit to the body. "The stomach is made to destroy chemicals," noted Zimmermann.

Finding a way to turn the compound into a drug that would not be immediately rejected or disintegrated by the body would not be easy. To enter the body, the molecule had to be stable in water without dissolving. To leave the body, the drug, having done its work, had to penetrate the portal vein, which goes through the liver, where the remnants could then be broken down for excretion. And these qualities had to be instilled in the molecule without disrupting its selectivity for the Abl kinase, the strength with which it bound its target, and its ability to kill cancer cells.

Although the formulating work could be done in a laboratory, the next steps required giving the drug to animals. This was the domain of pharmacokinetics and pharmacodynamics. The development team had to know how it was absorbed into the bloodstream and distributed around the body. They had to know how the chemical would be metabolized by the body and eventually excreted. All of these factors were crucial to the compound's viability, just as important as its ability to block the kinase. If the chemical could not be safely eliminated from the body, then that was a problem because it could accrue in the kidneys or the liver and become toxic. Charting all of the biochemical affects on the body, the good ones and any bad ones, was also essential. Did the drug raise blood pressure? Did it cause diarrhea? Did it cause excessive thirst? Such measurements could only be obtained by giving the drug to animals.

Within Ciba-Geigy, each department had its responsibilities and its boundaries. The wall separating the discovery team, which created new drug candidates, and the development team, which took over once the lead candidate had emerged, was tall and thick. "Once you have given your candidate compound to development . . . you have no say and no power over this compound anymore," said Matter. "You could talk and talk and talk, but they will do whatever they believe has to be done." As soon as the candidate compound was handed over for toxicology testing, Matter and his group—Lydon, Zimmermann, Buchdunger, and the rest—had no control over what happened.

From Matter's point of view, that separation was problematic. The development team, he felt, considered the early researchers to be "just fools and clowns," whereas they, the developers, had the real know-how about making new drugs. They also tended to be very conservative, understandably averse to taking risks with an unknown chemical. These two traits seemed to enhance one another, creating an endless loop of caution.

That caution surfaced very soon after the drug was passed from discovery to development. Medications get into the body through a few different delivery systems. A drug can be made into an intravenous formulation, injected through a needle into a vein; a subcutaneous formulation, injected under the skin; or an oral formulation, a tablet or liquid given by mouth.

Matter, Lydon, Druker, and the rest of the discovery team had been hoping to turn CGP-57148B into a pill. The preclinical observations of how the drug worked in cells indicated that the drug would need to be given fairly frequently to continuously shut down the cancer-driving kinase. In Druker's earliest experiments, the kinase had to be shut down for sixteen hours at a time for the cancer cell to be killed, hinting that a daily dose might be required to treat CML in people. A pill would be far more convenient than an injection if the medicine had to be given every day. An infusion wasn't out of the question—plenty of cancer drugs were given as daily injections for a week or two, and sometimes the injections took hours—but having a bottle of capsules that could be taken at home was far preferable.

But Matter was quickly told by the development team that the pill approach wouldn't work. "We got our first spurious data . . . telling us this compound could not be given orally [because] it didn't have bioavailability," he said. In other words, the formulation required to turn the compound into a pill would render the drug inaccessible by the body. An oral formulation, Matter was informed, was out of the question.

Matter wasn't sure that conclusion was accurate. In his estimation, the development team wasn't giving the pill form a fair shot. He was pretty sure that the friction between the teams had somehow biased their result, but the tension also meant he had no way to contest it. He, Lydon, and Druker settled on an intravenous formulation instead. CGP-57148B would have to be turned into a substance that could be shot through a needle into the vein of a patient with CML. The drug would then circulate through the blood, entering cells and killing any that contained the Bcr/Abl tyrosine kinase. It wasn't the best case scenario, but at least the drug was in development. That was what mattered.

WHILE THE DEVELOPERS worked on the intravenous formulation, Druker prepared his first public presentation on the compound. He had just submitted the paper on the preclinical studies to a third journal, *Nature Medicine,* in mid-November, and was getting ready to speak at that year's annual meeting of the American Society of Hematology (ASH), held in late 1995 in Seattle, Washington. It is the largest professional meeting in the world for doctors who treat diseases of the blood, including the three liquid cancers (leukemia, lymphoma, and myeloma), and conference presentations are often pivotal moments in the careers of cancer researchers, notches on the wall of accomplishment.

"We have demonstrated specific killing of the Bcr-Abl expressing cells by CGP-57148B," read the study abstract. Druker talked the small audience through a set of slides chronicling the dramatic effects the compound had had on Bcr/Abl-positive cells, and the lack of change among unexposed cells containing the mutant kinase. "This compound may be useful in the treatment of CML and other BCR-ABL positive leukemias," the slides read. Druker was lead author on the presentation. Buchdunger, Lydon, and Grover Bagby, who'd recruited Druker to OHSU, were also credited.

The presentation was attended by about fifty people and didn't cause much of a stir. That was no surprise; cell-line studies are so preliminary that they are rarely the stuff of press releases and rapt crowds. A man named John Goldman was one of the few who approached Druker afterward. Goldman, an English oncologist who was famous for pioneering the use of bone marrow transplants for leukemia in Europe, had been alerted to Druker's work by Ciba-Geigy. Intrigued, Goldman followed Druker back to Portland after the meeting, sleeping in the ballroom of a Holiday Inn because all the hotels were full. The next day, Druker agreed to give Goldman some of the compound to test back at Hammersmith Hospital in London. When experiments in his own lab reproduced Druker's results exactly, Goldman knew this compound was special.

Yet the attention of this highly regarded hematologist did little to calm Druker's nerves. He knew the intravenous formulation had been made and that toxicology tests were finally beginning, and he was anxious for any news. He was also unsettled about the rumblings he was hearing from Ciba-Geigy about the clinical trial. Now, after the first encouraging planning meetings, he was being told that the company was considering conducting the initial human study solely at MD

Anderson Cancer Center, in Houston, Texas. "They weren't sure they wanted me to be involved," says Druker. In part, he understood. Interferon, the standard treatment at the time for CML, had been developed largely at MD Anderson, with a leukemia expert named Moshe Talpaz at the helm, and patients with CML flocked there because it was reputed to provide the best care. The top leukemia doctors in the world were there, and Ciba-Geigy wanted seasoned investigators for the phase I study. Druker had to admit that he was not the ideal person to lead a clinical trial. "How many CML patients do I have in my clinic? I have three." Yet he had been part of this compound's creation for years now, and he was one of the few medical oncologists in the world who had championed the idea of kinase inhibitors. "I had to fight for this, to make sure if we're going to run a clinical trial, I was going to be part of it," he said.

On this point, Druker wrestled as much with his own reaction as he did with the company. After all, wasn't the real importance that the drug be made? "If a drug works for people, should I care [about my role]?" Druker asked himself. "No, I shouldn't." But he couldn't let go. "This is my baby," said Druker. "This is what . . . I've staked my career on, and I want to be part of this clinical trial." He refused to give up, but knowing the company's lackluster commitment to developing the drug left him unsteady, uncertain of Ciba-Geigy's intentions when it came to his involvement. In the end, it was Lydon who brought Druker the reassurance he needed. "I always had confidence with Nick there, and with his guidance, that I'd be involved, and more importantly, that the compound was going to make it into clinical trials."

IN FEBRUARY 1996, Ciba-Geigy held a second meeting to discuss the clinical development of CGP-57148B. Even though the drug wasn't ready yet, preparations needed to begin long before the clinical trial. The team—those spearheading the development of new drugs from within the company together with the physicians who would conduct the tests—had numerous matters to settle. The company representatives included Lydon, Matter, and Ford, who was continuing to consult with Gratwohl. Druker was invited, bolstering his confidence that, at least for now, he was still in the running to be a so-called principal investigator, a clinical trial leader. John Goldman returned for the second planning meeting, too.

Druker also recruited Charles Sawyers, from the University of California—Los Angeles, to the team. Sawyers was an oncologist with expertise in leukemia. Sawyers and Druker had met on the oncology conference circuit years earlier and were kindred spirits. Both held medical degrees, both were irresistibly drawn to lab research, and both had zeroed in on leukemia as the focus of their work. In fact, Sawyers had cut his scientific teeth in the laboratory of Owen Witte, who, with David Baltimore, had been largely responsible for figuring out the link between *bcr/abl* and CML.

When Druker showed Sawyers his results from his first cell-line and marrow studies of CGP-57148B in early 1995, Sawyers was immediately on board with Druker's desire to ask the drug company to run a clinical trial. "Why wouldn't we?" he said to Druker, awestruck by the data. Sawyers knew that the kinase program was a peripheral project that held little interest for the company. In part, Druker's request for Sawyers to join the clinical trial planning team was to show the company that, together with John Goldman, they could recruit enough CML patients for a phase I clinical trial.

In designing a clinical trial protocol, the team had to decide whether the compound should be tested only for CML or also other types of cancer; which stages of CML would be included; where the studies would be conducted; how many investigators to include; how many patients to enroll; and on and on. There would be reams of paperwork to prepare, so although the molecule had not yet entered a single animal, there was no time to waste.

By the early spring of 1996, the intravenous formulation was ready, the team in Basel having added all the necessary attributes without destroying the compound's anti-kinase activity. Now the drug—as it had become—could advance to the next stage: animal testing. The rules of the FDA state that of the two species in which a new drug candidate had to be tested, one had to be a non-rodent. The intravenous formulation would be given to rats and dogs. A couple of different toxicology studies were planned. In the first one, the animals would be given the compound as a bolus injection; that is, the entire dose would be inserted into the bloodstream fairly rapidly. In the next study, the dose would be given as an infusion over three hours.

During the months when CGP-57148B was being converted from a drug candidate into an intravenous drug, a man named Peter Graf, a former director of pharmacokinetic studies at Ciba-Geigy, had continued to work on the oral formulation of CGP-57148B. Like Lydon and Matter, Graf knew that a pill version of the drug would be much better than an intravenous version. His additional research contradicted those first findings of which Matter had been so suspicious. Graf showed that the compound could be made into a soluble formulation, one that would remain intact in the body's watery ecology and be absorbed into the bloodstream upon being swallowed, rather than being digested and eliminated with other bodily waste. What others on the development team had concluded was impossible, Graf had accomplished with certainty. But with the intravenous form already in toxicology testing, the pill was stored away.

In mid-April 1996, Druker learned that the preclinical paper had finally been accepted for publication at *Nature Medicine.* On April 30, Druker's forty-first birthday, the report was published. The conclusion was the same one Druker had noted during his ASH presentation: "This compound may be useful in the treatment of *bcr-abl*-positive leukemia." It was the stock, neutral language of so many reports of experimental drugs, mild enough to be supported by the data, piquing interest without being overblown. At that moment, though, Druker was entirely uncertain about whether he would ever get a chance to find out what the compound could really do. The process, though moving along, seemed to be taking forever. And although animal testing had begun and study protocols had been discussed, Ciba-Geigy had still not given full, official clearance for an eventual clinical trial if the toxicology work went smoothly. For Druker, confident in the drug and ready to move forward, the pace was excruciating.

On May 13, 1996, Druker received a fax from John Ford, who was the liaison between the investigators on the outside and the company.

In two paragraphs, Ford delivered the news for which Druker had been so anxiously waiting. The compound had been discussed at a meeting with Ciba-Geigy's research board early that month. "Our proposals received a favourable reception and we are reasonably confident that full approval to proceed into clinical trials will be forthcoming," Ford wrote him. The next step was to complete the protocols for the human studies. Ford forecasted early July for those planning sessions. The FDA had been in touch with the company and was eager for a face-to-face meeting. It was happening.

A month later, Druker received a second fax from Ford inviting him to a meeting in Washington, DC, later that summer to discuss the next steps in the clinical development of CGP-57148B. The company was meeting with FDA representatives for a preliminary talk about filing an investigational new drug (IND) application. The FDA would tell Ciba-Geigy everything that was needed before a clinical trial could begin. The agency would also go over what clinical data were required for the drug to be considered for approval. Along with Druker, John Goldman, Charles Sawyers from UCLA, and Moshe Talpaz from MD Anderson, were also invited. These were the individuals whom Ciba-Geigy planned to involve in the phase I trial. They would be the principal investigators.

Suddenly the air had cleared. Everything was looking good. The first human study of CGP-57148B had finally come within reach. The company was on board, meeting with the FDA, finding out what would be needed for the road ahead. And Druker would be one of the investigators. The wait was over. The drug was going to be tested in CML patients.

Then, on July 9, 1996, the fax machine rang again with a third letter from Ford. The first toxicology report was in, and the news was not good. In a four-week study, a group of dogs had received daily infusions of CGP-57148B at doses of 6, 20, and 60 milligrams per kilogram. The infusions were given through a catheter into the jugular vein for 28 straight days, each infusion taking about 3 hours. Some of the dogs given the two higher doses had "massive necrotising thrombophlebitis starting at the tip of the catheter and extending into the lungs," Ford wrote. The drug had crystallized in the blood, resulting in clots. The problem had occurred in the first week of the study, and some of the dogs had died as a result. The toxicologists altered the infusion schedule a bit, but the problem remained.

The clots were puzzling. In the first toxicology study, with the bolus injection, rats and dogs tolerated the compound well. Ford and the toxicology team had no explanation. He suspected some "undiscovered technical problem," he told Druker. Regardless of the cause, though, the study would have to be repeated, resulting in a six-month delay in the schedule toward opening the first clinical trial.

Druker stared at the page. He could not believe what he was reading: "Given the seriousness of the toxicity and the time penalty which would be incurred in trying to rectify it, we feel inclined to abandon our plans to initiate trials with the intravenous formulation."

There was a bright side, though. Ford was hopeful that the program could continue with the oral formulation. Graf's work had now rescued the program. He reminded Druker that the intravenous approach was only used because the pharmaceutical development team had wrongly predicted that the compound would not be absorbed when taken orally. "Abandoning the intravenous trials will make all of our lives much more straightforward," Ford wrote. However, the oral formulation would now need to enter toxicology studies from the beginning. His revised estimate for when a human study would launch was March 1997. In the meantime, the company had cancelled its meeting with the FDA. The FDA required toxicology data in two species given the oral formulation before it would discuss the IND, even in a preliminary fashion.

Ford persisted in his encouraging tone. "Naturally, we are all very disappointed by these unexpected events but we sincerely believe they represent only a temporary set-back," Ford wrote, concluding with the cheerful news that CGP-57148B had been promoted to clinical development status by the company's board of development.

23

"NOT OVER MY DEAD BODY WILL THIS COMPOUND GO INTO MAN"

At nearly the same time that CGP-57148B was infiltrating the veins of laboratory animals, Ciba-Geigy and its rival from across the Rhine, Sandoz, were merging. The merger created the largest pharmaceutical company in the world, renamed Novartis. For Druker, the news wasn't without irony. It was Dana-Farber's agreement with Sandoz that had thwarted his work with Lydon years earlier. Now here he was in the midst of trying to develop the very kind of drug for which Sandoz had once barely mustered the minutest enthusiasm.

The mid-1990s was an era of pharmaceutical mergers that brought together other giants, including Glaxo and Wellcome, Pharmacia and Upjohn, and Roche and Boehringer Mannheim, among others. This spate of deals, which continued for several years, was the result of an industry-wide decrease in productivity in research and development and an onslaught of generic drugs. The Hatch-Waxman Act, passed into federal law in 1984, loosened restrictions for generic drug makers, allowing their products to enter the market on the basis of bioequivalence rather than clinical data. That is, as long as the manufacturer demonstrates that the generic drug has the same properties as the original, there is no need for costly clinical trials to test for safety and efficacy to earn FDA approval. The change made developing generics much cheaper and bringing them to market much easier. As patents expired, generics, usually exact replicas of the once-protected molecules, flooded the market, and revenues for brand-name manufacturers decreased. The moment a generic drug was approved, sales of the original faded fast, a moment known in the industry as the patent cliff.

The mergers were a response to these combined forces, a way for pharmaceutical companies to keep their profits up and their pipelines full. (Eventually, tactics to prevent such revenue loss would also include paying off generic drug makers to delay the introduction of their product onto the market and creating new indications for branded drugs to extend patent time. These practices, dubbed "evergreening," are generally considered shady, operating through legal loopholes, but can bring a company millions of dollars in additional revenue, even when the result is just a few extra months of brand-name exclusivity.)

Yet the promise of ramped-up research and development that inspired the merger of Ciba-Geigy and Sandoz did not extend to the kinase inhibitor program. When Novartis was launched, "everything ground to a halt," said Lydon. The initial toxicology report from the intravenous formulation of CGP-57148B had dampened the company's interest in the drug, and although other animal tests were now in progress, all of the clinical programs were being reexamined as a routine part of the merger. It was the usual big-company problem, Lydon said. "No one was a champion for the compound within Novartis." The company had little motive to push ahead a drug that would ultimately be given to very few people.

Frustrated by the slow pace at which this pharmaceutical behemoth was now moving, Lydon resigned shortly after the merger. He struck out on his own to start a small biotech company called Kinetix, which would be bought by Amgen a few years later, resulting in the clinical development of several kinase inhibitors. Elisabeth Buchdunger took over as the head of the biology department for Novartis's cancer research program.

Alongside Matter, Lydon had been the force behind moving CGP-57148B—renamed STI-571 under Novartis—toward clinical trials. Without his greatest industry ally, and in light of the first toxicology report, Druker was worried that Novartis would lose interest in the drug.

The *Nature Medicine* paper of April 1996 had garnered little attention. That spring, the headlines were focused on the promise of angiogenesis inhibitors. Pioneered largely by a scientist named Judah Folkman, angiogenesis inhibitors were supposed to kill tumors by cutting off their blood supply. James Watson (who, with Francis Crick, Maurice Wilkins, and Rosalind Franklin determined the helical structure of DNA) famously pronounced that cancer would be cured in just a few years via this new approach. Kinase inhibition barely scored a footnote in discussions of exciting treatments on the horizon. Druker received some calls to talk about his work, and inevitably the interviewers asked whether the drug would be moving into a clinical trial. "That was the logical question," said Druker. But for him, it was the toughest one. He didn't want to acknowledge that he didn't know if the drug would ever move forward, and he didn't want to risk antagonizing the company and thereby jeopardizing the compound by pointing fingers.

One journalist visited Druker for an in-person interview. Alexandra Hardy, a writer from Houston who'd recently relocated to Portland for her husband's work, had been assigned the story by the Associated Press. Hardy was lukewarm about doing the article. She'd covered medical breakthroughs before and knew they rarely lived up to the hype. "I did not think the drug was going anywhere," she said. Something else did catch her interest, though. "The way he interacted with his patients struck me more [than the research]," she said. Hardy, not generally fond of doctors, noticed immediately how respectful Druker was to those in his care. It left a lasting impression, even if the reason for her visit didn't.

Hardy's story was picked up by the *Oregonian*, and then the attention quickly faded—almost. A man named Bud Romine, who was suffering from CML, sent Druker a note saying that if the drug ever went into a clinical trial, he wanted to be the first patient. The article also brought more people with CML to Druker, now a more widely recognized expert, for treatment.

Meanwhile the toxicology studies for the pill formulation were continuing at Novartis. Rats were given a low, middle, or high dose of STI-571 for two studies of thirteen weeks each. Some of the animals experienced kidney problems in the first round, but in the second thirteen-week stretch, the problems disappeared. In some animals, sperm production slowed down. Rats given the highest dose had bloody or dark urine, swollen muzzles, and increased salivation. The animals were euthanized at the end of twenty-six weeks and their organs were weighed. In many of them, testicular weight dropped. Some liver problems arose but were not life threatening. Two of the rats given a high dose of the drug died as a result, but none of the rats in the low- or middle-dose groups died.

In some of the female rats, the drug caused problems with the development of follicles in the ovaries. Blood samples revealed that the drug accumulated in their bodies at a faster pace than in the male rats. In pregnant rats and rabbits given the drug, the fetuses were damaged. In lactating rats, the chemical moved from the blood into the mother's milk supply.

The studies continued for more than a year, far longer than the few months John Ford had predicted in his last fax to Druker, and much to the consternation of Alex Matter, who could only watch in silence, aghast. To investigate how STI-571 might affect the central nervous system, mice were given a single dose and observed for side effects such as tremors and impeded motor function. No problems occurred. To study cardiovascular toxicity, a group of rats was anesthetized and given a shot of the drug. The animals did not experience any heart trouble except a short-lived decrease in arterial blood pressure.

To study the gastrointestinal system, a group of mice was given varying doses of STI-571. Two hours later, the animals were fed a liquid containing a small amount of charcoal, which cannot be digested, to see if mice given a higher dose of the drug would eliminate the charcoal more slowly than those given a lower dose. If the drug was affecting the intestines, the charcoal would take longer to be excreted. The charcoal was passed by all the mice in about the same amount of time.

A group of beagles were given an oral dose of 60 milligrams per day for thirteen weeks to investigate any potential side effects. Some of the animals had severe diarrhea that resolved.

As the studies went on, Druker didn't hear much from Novartis. It wasn't that he was being ignored; the radio silence was just part of the standard procedure. The data were being recorded and would be assembled into a final toxicology report. And the studies took time. Each test took about three months, with a further three months needed for analysis, which would include sacrificing the animals so that a pathologist could conduct a full exam of the internal organs. Until the analysis was completed, the findings would not be distributed outside the company. Druker kept himself busy with research.

Finally, someone from the company contacted him with results. Once again, the news was not good. In a study of dogs given doses up to 600 milligrams per day, high doses of the drug had caused liver failure. In rats, even low doses had resulted in some liver damage. The drug was considered too dangerous for humans. One of the toxicology experts told Alex Matter, "Not over my dead body will this compound go into man."

To examine liver toxicity, the drug had been given in increasing doses for up to thirteen weeks. As the weeks passed, cells inside the animals' livers began dying off, and inside the bile duct, cells began growing at a pace that was faster than normal, often a precursor to cancer. At regular, frequent intervals, the toxicologists checked the level of enzymes in the liver. Heightened levels of enzymes indicated liver damage. But when they noticed that liver enzymes were elevated in the animals, they continued to give the drugs. By the end of the study, two or three months later, the dogs had liver failure. Four weeks after the study, some of the dogs still showed signs of liver trouble.

Druker took issue with the rationale behind their approach. "If you know the dogs' liver enzymes were elevated, wouldn't you stop?" he explained. "If I was giving this to people, I'd stop the drug." Unschooled as he was in running a clinical trial, Druker knew that administering experimental cancer drugs followed the same principles as giving highly toxic chemotherapy. When a dose is too high for a patient to tolerate, it's lowered. When the side effects are too much to bear, the treatment is halted. Cancer patients aren't force-fed medications until their livers fail. Rather, the functioning of all organs is carefully monitored throughout treatment. "We know what to look for, we know when to hold the drugs," Druker said. In a clinical trial, patients are monitored even more frequently than in a regular clinic.

Further, the liver problems occurred most often when the dose was 2,000 milligrams per day or more. Druker knew from his preclinical studies that the ideal dose of the drug would probably end up being much lower. So he wasn't bothered by liver toxicity. The point of doing the toxicology test was to see how the drug affected the liver, not to kill the animals by giving them excessive amounts of it. He knew that if he saw elevated liver enzyme levels in people with CML, he would stop giving them the drug. There were concrete warning signs that any oncologist would heed, so the risk of liver failure was extremely minimal.

The company didn't see it that way. "Their view was, nope, we're going to have to go into monkey studies, and that will take another couple of years," Druker recalled. He was told that once the results from the monkey studies were in, Novartis would then decide whether to launch a clinical trial.

The company feared that the FDA would see the liver data from the canine studies and forbid the clinical trial out of safety concerns. In fact, a member of Novartis's oncology division had previously worked at the FDA, and he insisted that the drug would never make it through their review of toxicology data. To Druker, that concern seemed overblown. After all, the standard medication for CML at the time, interferon, did not work for many patients, and those whom it did help often suffered terrible side effects. Bone marrow transplantation, the only true cure for CML, was an option for only a fraction of CML patients, who first had to survive the harrowing and dangerous procedure before they could be assured of living CML-free. With such a paucity of therapeutic options, would the FDA really thwart a trial of a new drug because it might cause liver toxicity, an easily identifiable problem that could be stopped long before it became life threatening? "That just didn't make much sense to me," said Druker. Plus, a colleague later told Druker, dogs have particularly sensitive livers, and so are far more prone to side effects in that organ than are other lab animals, or humans.

Inside Novartis, concerns about toxicity seemed prudent and warranted. Having seen a severely toxic side effect in an animal, could the company really move forward with treating patients? "If we do so and then a patient dies, we will be in deep, deep trouble," said Zimmermann. "The company can lose its reputation when you make a mistake."

That concern was far from hypothetical. The damage seen in the dogs could have been a sign of what was to come in CML patients. If it were, patients were going to die from the drug. The shadow of thalidomide, the drug used by pregnant women in the 1950s that turned out to severely deform the limbs and organs of fetuses, with 40 percent of "thalidomide babies" dying within their first year of life, still hung over the pharmaceutical industry. Prior to its use in people, the drug had been given to rodents only, and the disaster was one of the reasons the FDA started requiring animal testing on multiple species. (Today, thalidomide is used in the treatment of multiple myeloma, another type of blood cancer, but that benefit was discovered only decades after this initial debacle.) The possibility that STI-571 could be such a calamity was very real. "Then you'd have to go back and say, 'Why did we risk it?'" said Zimmermann.

With Lydon gone, Druker knew that it was up to him to fight for the drug. "Brian was extremely focused on his patients," says Lydon. "He felt it was a pretty hopeless and depressing situation in the clinic at the time for CML, basically watching them progress and not much you could do about that." Druker understood the need to be cautious, but only to an extent. Unlike the industry representatives hesitating in Basel, he saw dying cancer patients every day. He knew that alongside the safety concerns, the marketing hesitations lingered. Was the company being cautious, or was it hoping to use the results as an excuse to not move the drug into costly clinical trials?

He also knew that the potential impact of this drug stretched far beyond CML. If STI-571 proved that the principle behind kinase inhibition—that cancer could be stopped by selectively targeting a single haywire protein—was an effective way to treat cancer, then other such drugs could follow for other cancers. STI-571 was the first drug targeted against an abnormal protein resulting from a genetic mutation. If it worked, how would that change the direction of cancer care? How would that change our understanding of the underlying cause of cancer? What powerful medications for cancer or other serious diseases might then be possible? Druker knew his vision might seem grandiose or, at the very least, decades away from becoming a reality. But the current chemotherapy had evolved over at least fifty years. Maybe this drug was the start of the next fifty. As unlikely a dream as it was, it didn't seem completely impossible.

At some point, the decision makers at Novartis told Matter that if his discovery team could find another use for the drug, then they would consider a clinical trial. Matter and the team knew that STI-571 had shown activity against two other tyrosine kinases, PDGFR and Kit, when the compound was first developed. Now it was time to see if those targets might have any clinical usefulness. Matter turned once again to Chuck Stiles at Dana-Farber, who quickly showed that the drug was active against glioblastoma multiforme, a type of brain tumor that expresses PDGFR, in rats. Years earlier, Lydon, Buchdunger, Zimmermann, and Matter had fretted about whether to try to eliminate the activity that STI-571 had against PDGFR, wondering if it might make the compound a stronger inhibitor of Bcr/Abl. Now they could only be thankful that they'd left well enough alone.

Their requirement satisfied, the company still resisted initiating a clinical trial. Druker called colleagues who he thought could weigh in on the value of the toxicology studies. He suggested to Novartis executives that they call the FDA to ask whether their toxicity data were sufficient to submit their request to run a clinical trial. "We can't do that, we don't have permission," Druker was told. "What if I talk to the FDA?" he asked. "You can't do that," he was told. "So," said Druker, "I did it anyway."

At Novartis, the monkey studies—the seventh toxicology test—commenced. To further explore the liver toxicity and other potential problems, a group of cynomolgus monkeys, macaques found in southeastern Asia, Borneo, and the Philippines, were given varying doses of the drug per day for thirteen weeks. None of the monkeys died as a result of the drug. The animals receiving the high dose vomited or had diarrhea, and four from this group lost weight. In several of the monkeys given the high dose, their gums turned pale. Many of them had changes in their red blood cell and white blood cell counts, but all returned to normal after the drug was stopped.

Matter, who had been made head of oncology research when Sandoz and Ciba-Geigy merged, was fed up with the toxicology testing and could not remain silent any longer. He had to do something to put an end to what he considered a ridiculous course of action. He was known for his hot temper—"It was my profession to rant, to wail and to remonstrate, at every level," he said. When the monkey studies started, Matter met with the top management team at Novartis and "scolded them for half an hour," as he put it.

When Matter was done, Pierre Douaze, the temporary CEO following the merger, stood up. His response, Matter recalled, was to "congratulate himself for allowing such language in a meeting with top management." For all of Matter's ranting, STI-571 remained stuck in toxicology limbo.

Matter's insistence that the company move the compound forward was leaving its mark, though, however faint. "They knew that I was not going to cave in," he said. But his efforts did little toward grinding the clinical wheels into motion. Even as head of oncology for what was then the largest pharmaceutical company in the world, Matter was almost powerless. He was fighting a fight that has dogged drug development for ages. Companies want to know whether a drug is toxic because they don't want to give a dangerous drug to people. Clinicians insist that they will be vigilant and that it's not right for a company to withhold a drug that could benefit patients immediately.

In the meantime, Druker had gone ahead and called the FDA and was told that Novartis had accrued plenty of safety data to warrant review for a clinical trial. "You have way more than most companies have for these drugs," Druker was told by a toxicologist at the agency. He relayed the message to executives at Novartis, immediately drawing their ire. "It didn't advance anything," said Druker. "All it did was get them madder at me."

Druker was running out of steam. If the people at Novartis would not even listen to a message directly from the FDA, how could he possibly persuade them to launch a clinical trial? A patient of his who'd run out of treatment options begged for a sample of the compound from Druker's lab supply. A few weeks later, the patient died. Desperate to find a way to get this drug to the CML patients who needed it, Druker called Nick Lydon to see if he had any ideas. He did.

Lydon suggested that Druker write a letter to Alex Matter with the forceful request to make a decision: Either go into clinical trials or license out the drug. Novartis could easily sell the drug to another pharmaceutical company. Most of the large industry players weren't interested in tyrosine kinase inhibition, but the number of small biotechs was increasing. These companies were dedicated to the research and development of innovative new drugs such as biologic-based monoclonal antibodies and small-molecule inhibitors, drugs that were little enough to slip through the cell membrane and attack the cancer-inducing mechanisms from within. A modestly sized company could run on a single promising compound, trying to generate investors with preclinical data, either waiting to hit the big time with a drug that actually worked or hoping the research would garner an offer from a larger, wealthier pharmaceutical company. Any of these companies might be glad for a chance to buy an experimental drug that already had so much data behind it. Lydon had even suggested to Druker that his company could buy the rights from Novartis. They knew Matter was on their side, but he was the person to address in this formal way. Druker should explain to Matter that if the monkey studies show toxicity, all it's going to tell them is what dose to watch for in clinical trials. If they don't show toxicity, Novartis will have wasted millions of dollars in two years. Druker had patients who needed it immediately.

Druker followed Lydon's advice. He told Matter that investigators who'd been through the IND process and representatives from the FDA itself all thought the compound deserved a chance to be studied in a phase I clinical trial, the first stage of human studies. He told him that the ongoing thirteen-week study in primates was a waste of time and resources. The company already knew about the liver toxicity that resulted from continued dosing of the drug. If the monkey test turned out negative, careful liver monitoring would still be required in the human study anyway. If the monkey test turned out positive, then a potentially useful drug would probably be dropped without testing in humans. There were already enough toxicology data to submit the compound for IND approval, Druker insisted.

He also addressed a concern expressed by the toxicology team at Novartis that giving the compound for more than four weeks would be unethical. That perspective was based on a strict interpretation of an FDA rule stating that an experimental drug could be administered in a phase I trial for a third of the duration of the toxicology studies. The animal tests had lasted twelve or thirteen weeks each, so Novartis insisted that the phase I study could give the drug to patients for only four weeks at a time.

In Druker's estimation, that viewpoint simply did not hold water. The duration of treatment in actual patients depended entirely on how the benefits weighed against the risks. CML patients had a uniformly poor prognosis, and those who would enroll in the study had no other treatment options. Give them the drug, said Druker. If the patients benefit from it and tolerate it, then there would be no ethical reason to stop treatment. Monitoring for liver toxicity was not complicated. Blood tests would reveal any problem, and the investigators could even do liver biopsies if necessary. "It is time to make a decision," Druker wrote. "Give the drug a chance." If Novartis did not want to take the compound further, then the company should license it out. He told Matter that Nick Lydon's biotech company would probably be interested.

Druker sent the letter to Matter. Strengthened in his resolve by Druker's words and dedication, Matter took a deep breath and dived back in, hoping to persuade the company to finally make the so-called go/no-go decision.

MATTER KNEW THAT if he didn't fight for the drug, Novartis would shelve it. "We had to champion these programs, or they would die." The persistent pleading from clinicians on the outside gave Matter the confidence to keep fighting. The letter from Druker and all the other voices in support of the drug gave Matter the boost he needed. "Without [everyone] covering my back, I don't know whether I would have had the stamina to see this through," Matter said. It was Druker's words he would hear in his head—*give the drug a chance*—when he found himself, month after month, pleading STI-571's case.

Some weeks after his rant to the executives, Matter bumped into Daniel Vasella in the hallway. Vasella, 42, had replaced Douaze as the permanent CEO and chairman of Novartis, and was still getting to know his employees. Born and raised in Switzerland, Vasella had been a doctor of internal and psychosomatic medicine for eight years when, in the mid-1980s, he started wondering what it would be like to work in the pharmaceutical industry. His wife's uncle was the head of Sandoz, and Vasella, curious about the business of medicine, peppered him with questions. In 1987, the new head of Sandoz offered Vasella a job in marketing, though it required working out of the company's New Jersey headquarters. He and his family headed to the United States.

That phase of his career would give Vasella his first taste of patient activism. Sandoz had just launched a drug called Sandostatin, or octreotide, designed to treat a rare tumor that caused severe diarrhea. At the same time, the AIDS crisis was at its peak, and patients, who often suffered from diarrhea, learned that there was a new drug that could help. "They wanted access, and we had to manage it in a way that was sustainable and legal," recalled Vasella. It was his first confrontation with extremely angry, activist patients, and also his first encounter with a drug created for a single, rare condition that turned out to have multiple applications. The success of the drug was gratifying, and its lessons resonated as Vasella ascended the industry ranks.

Vasella and Matter had met early in the merger process, as part of efforts to integrate the two companies' teams. "He was the most angry person I ever met," Vasella recalled. Matter's first impression of Vasella was that he was a suave businessman whose calm disposition stood in stark contrast to his own more combative temperament. "He was a picture of a hero," recalled Matter. "He was very educated, elegant, an eloquent new presence."

In that chance hallway encounter in the middle of 1997, Matter immediately started venting his frustrations about STI-571. "He said he had a promising drug, but we couldn't get it into clinics," Vasella recalled. Still familiarizing himself with Novartis's drug portfolio, Vasella knew little about STI-571 since he'd come up through Sandoz, not Ciba-Geigy. He didn't know about the basic research that connected the Philadelphia chromosome with CML, and he knew very little about the rationale behind kinase inhibition, though he knew Matter had fostered the initiative within the company. He didn't know about the cell-line studies or the endless toxicology studies about which Matter was giving him an earful. All he knew at the time was that the person standing before him was insistent about the potential value of STI-571. Vasella asked Matter if he was certain about this compound. "I really believe it," Matter told him. "Okay," Vasella said, "then we're going to do it."

Vasella's response was, in part, simply that of a CEO handling a hotheaded employee. "I was basically, I think, coping with it in a way which allowed him to believe that I could maybe listen," said Vasella. But it was also the first time that the drug was brought to his full attention, and Vasella found he couldn't ignore it.

As Vasella began investigating the kinase inhibition project, he became increasingly aware of the skepticism surrounding the development of STI-571. It was the prevailing mood at the company, he noted, not traceable to one person or team. "You will never know who it is because it will be some of the researchers, some of the developers," says Vasella, "[and] the marketers who think, 'Well, this will never make any profits.'" Whether from his experience with Sandostatin or solely on the merits of the STI-571 data thus far, Vasella was persuaded to consider moving the drug into a clinical trial.

In August 1997, Druker received a confidential letter from Matter. As Matter put it, Druker's letter had given him the motivation to go straight to the top. Energized by Druker's uncompromised belief in the drug and his own convictions, Matter had brought the compound and the flagging clinical development to the attention of the CEO, the global heads of research and clinical research, and the head of clinical oncology research. As far as he could tell, they were willing to listen, and he hoped to have more news in a couple of weeks. "For the time being, I would like to express my feelings of gratitude for your dedication to this project," Matter concluded.

Four months passed before Druker finally heard from Matter again. Matter divulged that a major battle had been raging over STI-571 among different departments at Novartis. He did not describe the details, but Druker knew what the problems were. It wasn't only the persistent concerns about the potential side effects of the drug that had resulted in the toxicology team taking a stance against continued development. In addition, the marketing team was still insisting that the drug would cost the company too much money. Using an estimate of the patient population, the assumed duration of treatment, and the extent to which the drug could penetrate the market—some doctors, they assumed, would likely continue to use the current standard therapy—marketing had come up with a projected total sales figure of about $100 million, an amount that was, as Matter put it, "totally beside the point."

There was no way the marketing team could have projected accurate sales data because kinase inhibition had never been attempted. It was a completely new kind of medicine. No one knew what the duration of treatment would be, the exact patient population for which it would be appropriate, or how quickly doctors would adopt the drug into their standard of care. These calculations were simply outside their scope of reference because this drug was an entirely new way of treating cancer.

Matter was happy to report to Druker that at last the evidence and science behind the drug had prevailed over the business hesitations. A pre-IND meeting had been scheduled with the FDA for sometime in the next two months. "So, with delays, we are still alive," Matter wrote.

Finally, on December 23, 1997, Druker received a letter from John Ford, now the senior clinical research physician for Novartis, via Federal Express. The clinical group had finally managed to persuade the toxicology group to at least discuss the possibility of starting a clinical trial with the FDA. The process, Ford couldn't help but mention, was likely advanced by the resignation of the head of the toxicology group. Vasella had approved moving the drug into clinical trials. The pre-IND meeting had been requested, and would likely happen in January or February. If the FDA gave the program its okay, then a clinical trial would begin around May. If the FDA requested changes to the proposed protocol, the trial could begin in June. If the FDA requested additional toxicity data, then the trial would have to wait until the last quarter of the year. Whichever option panned out, STI-571 would be in trials that year.

MEANWHILE, THE MONKEY studies continued. In the high-dose group, the animals vomited or had diarrhea. Many of them lost their appetite and became dehydrated. One of the female monkeys in the high-dose group died after losing 15 percent of her body weight, but it wasn't clear to the researchers that her illness was due to the drug. Many of the animals had various forms of *Plasmodium*, a malarial parasite common in monkeys bred in Africa and Asia. The parasite was the more likely culprit behind the changes in red blood cell counts, but the drug may have allowed the parasite to proliferate. The lowest dose of the drug given to the monkeys did not cause any problems in their blood or gastrointestinal systems. All of the male monkeys had testicular changes as a result of the drug. The company concluded that the drug was "tolerated well by the animals with minimal toxicity," according to the report eventually submitted to the FDA. The high dose, however, would cause "definitive toxicological/pharmacological signs."

To examine the effects of a particularly high dose of the drug, a group of monkeys received 1,800 milligrams per day for six days before an excess of toxicity led the scientists to reduce it to 1,200 milligrams per day for the remainder of the study. One of the animals had to be euthanized after it developed severe kidney disease.

All told, STI-571 was studied for thirty-nine weeks in rats, thirty-nine weeks in one group of monkeys, thirteen weeks in another group of monkeys, thirty-nine weeks or more in mice, at least twenty-six weeks in rats, and at least two weeks in rabbits. It was studied in rat fetuses, cells from Chinese hamster ovaries, and bone marrow from rats. It was mixed with *Salmonella* and *E. coli* in tests designed to show whether the drug would cause genetic mutations. A special additional study to further investigate toxicity in dogs concluded that, although liver problems remained a concern, the drug was safe.

Theoretically, the early indications of the potential efficacy of STI-571 should have been enough to launch a study within half a year after Druker reported his cell-line and marrow data to Matter and his team at Ciba-Geigy. The mandatory toxicology studies could have taken as little as three to six months. In principal, pharmaceutical companies can move ahead incredibly quickly if all signs point toward go. "Let's assume everything has been done to go into humans and you have a hypothesis [that] you believe in, and you have a patient population who is potentially going to profit from it. It's only the decision [that remains]," said Vasella. "It shouldn't take months. It should not." And yet STI-571 lingered in its preclinical state for nearly three years after the completion of the cell-line studies, and for about two years after the initial toxicology studies and the business merger. The kinase inhibition program had started in 1984. The lead compound had been synthesized by 1990. Seven years later, the phase I study had yet to begin.

PART 3

Human Trials

1998–2001

Behind every new drug entering clinical testing lies a principle justifying why it should work, a mechanism proved by similar drugs already on the market. But when a new kind of medication enters trials, that proof of principle is nonexistent. Testing the drug is the only way to test the principle.

To prove the principle that cancer could be treated by targeting the abnormality driving it, STI-571 had to be given to patients with CML. Everyone involved—the investigators, the company executives, and the patients—had to accept the risks of testing this novel drug. As solid as the rationale behind the design was, proof could come only from evidence. Success was far from guaranteed.

24

THE QUICKEST ANSWER

The phase I trial of STI-571 was a first for kinase inhibition, but also for Druker: He'd never experimented on patients before. It was a moment he'd envisioned for years, picturing how he would be with the patients and what he needed to explain to them. Their lives would be in his hands, and his job was not only to keep them safe, but also to make sure they knew what they were in for. He knew that giving them the best possible care meant not raising their hopes that this untested drug would do any good.

Patients enrolling in phase I clinical trials are always informed that the goal of the study is not to help them but to help advance medicine. This phase, the first of three that are required for FDA approval of a new drug, is also known as a dose-finding study. Patients are started at an extremely low dose of the experimental medication, and that amount is gradually increased. The idea is to find out how much of the drug a patient can safely tolerate, based on the principle that the more drug that is given, the more effective it will be. The highest dose at which the drug can be given safely is called the "maximum tolerated dose." Identifying that dose is the goal of a phase I study. These are safety trials, designed to ensure that the drug is not too dangerous for human consumption.

For this reason, the cancer patients who enroll in phase I studies for new drugs are usually those who have run out of treatment options. A newly diagnosed patient who has not yet received the standard treatment for their disease would not be asked to try an experimental drug. Rather, phase I trials are populated by people whose cancer has progressed after trying every other treatment. These patients must choose between accepting palliative care to ease the pain until they die or trying an experimental drug, if an appropriate one exists, in the name of helping science. When they enroll in phase I trials, cancer patients sign consent forms acknowledging their understanding of what the study is offering. They aren't supposed to expect the drug to work or to hold out hope that it will.

But the design of the STI-571 trial was different from most. Phase I studies are always small, and this one followed suit: The study would enroll about thirty patients and would last six to twelve months. The distinguishing feature of this phase I trial was the decision of which CML patients would be permitted to enroll.

Despite the endless delays, many features of the phase I protocol had already been decided. From the earliest meetings at Ciba-Geigy in 1996, the team agreed that the trial would be restricted to CML only. That approach was a departure from the typical phase I trial design, in which new drugs are tested on multiple cancer types at once, in case the agent works for many malignancies, or for one tumor type but not another. That approach makes sense for drugs that attack cancer in a general way, working against mechanisms that are, for example, present in normal cells but heightened in cancerous ones. Often, it's not until the later phases of clinical trials that enrollment is restricted to one specific cancer type, the one the company deems most likely to win FDA approval based on the broader phase I investigation.

But this was a new kind of drug. Aside from tamoxifen, which targeted the higher levels of estrogen present in some breast cancers, cancer drugs weren't aimed at specific peculiarities of specific cancers. With the kinase inhibitor, restricting testing to only CML made sense because the drug blocked a protein linked to a genetic mutation specific to this cancer type.

Next they had to figure out what stage of CML to test. Focusing on blast crisis, the final stage of the disease, would have been in keeping with the traditional phase I structure, but the company didn't want to do that. By the spring of 1996, when Ciba-Geigy held the second trial planning meeting with outside oncologists, Ford and the other executives had boldly decided to test the drug not on the most advanced patients, nor on those in the middle, accelerated stage, but rather on those in earlier stages of the disease. "My initial response was, 'Can you do that?'" said Druker, who knew that experimental drugs were tested on only the sickest cancer patients as a matter of medical ethics. Yes, he was assured, they could.

That change wasn't the most radical one brought to the table. Ford and Gratwohl had even discussed testing the drug on newly diagnosed CML patients as a one-month treatment before they went on to receive the standard therapy, although that idea was quickly dismissed. "In 1996, no one would have considered that," said Druker. The FDA and the medical establishment consider it unethical to give an unproven drug to patients who have not tried the current medication. Enrolling earlier-stage patients whose disease had progressed with interferon was different enough. "We already knew we were doing something unusual in a clinical trial," said Druker, who took a stance against enrolling newly diagnosed patients.

When the time finally came to open the trial in 1998, Novartis maintained those early decisions. The study would enroll CML patients only, but those in blast crisis—the very sickest—were not allowed to join. The study would be open only to patients who'd failed interferon but were still in the early stages of the disease.

At first glance, it could appear that Novartis designed the phase I trial to maximize the possibility of benefiting the CML patients who enrolled in the trial. Didn't limiting the trial to only this disease signal the company's understanding of the CML-specific nature of the drug, and Novartis's commitment to helping patients in need? And enrolling patients who were in better shape than those in the final stages of the disease gave the drug more time to work before patients died.

But these weren't the reasons, or at least not the main ones, behind the phase I protocol choices. Rather, the design was aimed at getting the earliest possible answer about whether the drug worked. "They were very concerned about committing to a trial that didn't have a clear and quick readout," Druker said of the trial planners at the company. Yes, the decision to limit the study to CML only was based on the fact that the Bcr/Abl kinase was so prevalent in this cancer and had not been found in any others. But the reason why the study was limited to CML wasn't so that patients who needed the drug could get it, but because that was the only way to determine the drug's potential efficacy. If the drug was not tolerated in a study with multiple cancer types, then clinicians might argue for a phase I study limited to CML only. But if the drug was not tolerated in a study of CML patients only, then it could be discontinued. And even though the objective of the phase I study was to find the maximum tolerated dose, the company wanted to know as soon as possible what the drug did in patients with CML.

Opening the trial to patients in the earlier stages of CML, rather than restricting it to the sickest patients, followed similar logic. Ford and others at the company thought that blast crisis patients would never respond to the drug. Druker had seen the drug work on cells from blast crisis CML and so thought there was a chance the drug would work on these end-stage patients, but the company didn't want to risk a vague study outcome that would leave the drug in limbo.

Testing STI-571 on the healthiest CML patients in phase I would give the clearest indicator about whether or not the drug worked. If the drug was poorly tolerated among chronic-phase patients or showed absolutely no anti-CML activity in those patients, then perhaps the drug could be quietly shelved. A phase III trial—the costliest of the three clinical studies—for a rare disease with a limited market could be safely avoided.

Even at phase I, the investment was formidable. Novartis would have to cover almost all the costs of the trial: the medication, the battery of tests necessary to monitor its safety, biopsies, and any other procedure related to the care of each patient's CML. There were expenses for recording test results and all the other data, for analyzing blood samples, for treating side effects. The costs were no different from those of developing any other new drug, though making this one was incredibly labor-intensive, requiring an arduous twelve-step process involving hazardous materials; producing just the single kilogram that was used for phase I took months. The major difference here was that, unlike industry sponsors for most phase I trials, the company was still not entirely committed to bringing this drug to market.

Druker and the other investigators were completely on board with the plan. They knew that this unusual phase I study design was the best way to prove whether the principle of kinase inhibition was valid. "Phase I is for safety, we knew that," says Druker, "but the reality is we also wanted to plan a study to see if this thing worked or not." The unconventional approach would enable the clinicians to see if the drug could be tolerated, indicating that it was, most likely, selective for the cancer-inducing kinase. And it would reveal, at least in a preliminary fashion, whether the drug affected the cancer.

The design also enabled them to investigate whether the concept of kinase inhibition was a viable approach for treating cancer. If the drug showed promise, then the results of the phase I study would set the stage for more profitable kinase drugs to enter the development pipeline. If the drug didn't work, then Novartis would have minimized its losses by conducting just this single, small phase I study before canceling the kinase program altogether. The study had been designed to end the project as quickly as possible, if an ending was warranted.

THERE WAS MORE to decide than just who would take the drug. The team also had to figure out how the drug would be given. Even though the doses were experimental, the frequency with which the medicine was taken could have a huge hand in how well patients tolerated it and whether it affected the cancer. Clinicians from outside the company were pretty certain that the drug would have to be taken every day. Druker's lab experiments had led him to believe that the kinase would have to be shut down for at least 16 continuous hours; this understanding had provided the impetus for the oral formulation years earlier. This cycle meant patients had to take a pill each day.

Novartis was pushing for a break in the regimen. That was how most chemotherapy regimens were done; treatments were frequent but not every day. Fearing toxicity, people at the company thought that a recovery time should be built into the trial. It took several rounds of discussions to settle on the once-daily approach. As everyone came to terms with the notion of daily treatment, the availability of the pill formulation—thanks to Peter Graf's persistence years earlier—now seemed all the more fortunate. Phase I patients would need to be at the clinics regularly, but they would not have to sit for hours in chemotherapy chairs with an experimental fluid dripping into their veins.

As for how long the trial would last, the plan was for the clinicians to gather data for about six months, following the mold of typical phase I cancer drug trials. However, those six months included the treatment plus a few months of follow-up time to see how patients did afterward. Everyone assumed that the treatment itself would last just a month or two. Again, the history of chemotherapy was shaping their thinking. Most cancer treatments lasted a month, and then they were over. With chemotherapy, the goal was to kill all the cancer cells with the poison in a short amount of time, and then stop before the poison did irreparable harm to the rest of the body. Whether in a standard clinical setting or on a clinical trial, patients didn't remain on chemotherapy indefinitely. That wasn't how such medications worked. Nobody imagined this drug would be any different.

25

GETTING TO 200 MILLIGRAMS

As the trial finally got under way, Druker remembered Bud Romine's letter from two years earlier asking to be his first patient if the medicine he'd read about in the *Oregonian* was ever tested on people. Romine, who lived in Tillamook, Oregon, was 64 when he was diagnosed with CML in 1994 and told he had about three years to live. Two years later, he saw the front-page article on Druker in the Portland daily newspaper and wanted to try the drug that had so effectively killed the malignant cells in bone marrow samples. When it was time to start enrolling patients, Druker remembered the bold request.

The study opened at three sites: OHSU, where Druker was the lead investigator; UCLA, headed by Charles Sawyers; and MD Anderson Cancer Center, in Houston, Texas, headed by Moshe Talpaz. Talpaz was famous in the oncology community for his work developing interferon, which, despite its harshness, was an extraordinary advance for CML. Of the three, he had the most experience with clinical trials. In fact, he was the only one who had led clinical trials; the company had first considered asking Talpaz to run the trial entirely, back when Druker had feared he'd be omitted. Left out of the equation was John Goldman, whose group at Hammersmith Hospital in London had been the first to confirm Druker's 1996 results and had been asking Novartis to be part of the trials ever since. Wanting to study this wholly new drug in the most controlled way possible, the company had opted to restrict the phase I study to the United States.

Starting in June 1998, each site would enroll one patient per month, for about ten months, for a total of about thirty patients. Druker, Sawyers, and Talpaz knew that the low starting dose—25 milligrams—would not have any effect. But starting low was essential to preventing dangerous toxicities. Month by month for up to a year, they would raise the dose higher and higher, looking for a change in blood counts and keeping vigilant for side effects.

The pills were small and orange, the whitish powder that Jürg Zimmermann had created held inside a shell made of gelatin and dyes. Upon entering the body, the shell dissolved, releasing crystals of STI-571. Until now, those molecules had been released only into the bodies of rats, mice, dogs, rabbits, and monkeys. Every clinical trial investigator knows that animal studies are often poor predictors for what a drug will do to a human being and how a human body will handle a drug. So regardless of the years of toxicity testing and the endless fights with Novartis to test the drug in people, Druker, Sawyers, and Talpaz still proceeded with trepidation. They stood at the crossroads of the theoretical power of hitting a single kinase and the unknown reality of what would happen when this foreign chemical was unleashed in the body. Would it really block just the one Bcr/Abl kinase and leave the person unharmed? Or was there a chance it would shut down multiple kinases, an event that, it was not melodramatic to think, could kill someone. "What if it blocks ATP binding of every single kinase in the body?" said Sawyers, recalling the concern at that time. "That's going to be one sick person."

In Portland, Bud Romine barely hesitated before swallowing his 25 milligrams. The same for the minister from Bakersfield, California, who was Charles Sawyers's first enrollee. They would remain in the clinic for eight hours, under careful surveillance for any signs of potentially dangerous side effects, and then return the following morning. In Houston, it was the same.

Enrollment required a commitment. The phase I patients had to live near the study site for about three months. In Portland, Los Angeles, and Houston, patients, often accompanied by their spouses, rented homes, temporarily leaving their lives behind for a chance to try this new drug. They also had to agree to close medical scrutiny, including blood tests three times a week, periodic bone marrow biopsies, and other monitoring. For each patient, the investigators had to record white blood cell counts, red blood cell counts, liver enzyme levels, kidney function, weight, temperature, blood pressure, and any other measures that might reveal a problem the drug was causing. Check-ups were at least once a week.

The investigators explained to the trial patients the various ways in which their response to the drug would be measured. First was the hematologic response rate, which showed changes in the number of red and white blood cells based on samples usually drawn from the vein at the crook of the elbow. A decrease in the number of white blood cells would indicate something positive. The number of platelets, the part of the blood responsible for clotting, can rise or fall as CML progresses, so a count moving toward the normal range was a good sign. A hematologic response was defined as a decrease in the white blood cell count by half. A complete hematologic response meant a patient had normal counts of white blood cells, platelets, and red blood cells, and no blast cells in the bloodstream.

Responses to STI-571 would also be measured in terms of a cytogenetic response. Cytogenetics examined the connection between genes and disease, and so a cytogenetic response meant a change at the genetic level as a result of the treatment. Using bone marrow biopsies (the same chromosome analysis tests that Janet Rowley had used in the 1970s), and FISH to a lesser extent, the investigators could examine how the drug affected the prevalence of the Philadelphia chromosome. Were there fewer cells containing this genetic abnormality following treatment with STI-571? When the bone marrow was biopsied at the time of diagnosis, usually 100 percent of the cells examined had this abnormality. A so-called major cytogenetic response meant the Philadelphia chromosome was now present in just 35 percent or less of the biopsied cells. A complete cytogenetic response meant the Philadelphia chromosome was entirely absent.

Druker knew that the starting dose of an experimental drug is always based on toxicity observed in animal studies. The first human dose would be one tenth of whatever dose caused animals to be sick in some way. But Druker also knew the drug level—technically different from drug dose—that was needed for the cancer to be killed. Concentration is measured in molars, a unit representing the amount of material held in a solution. Based on his studies in CML cells, Druker predicted that a human bloodstream would need to contain a minimum of one micromolar of STI-571 for the drug to have any anticancer activity. At less than that, the drug would be too dilute to do any good.

However, the dosage needed to achieve those concentrations—one micromolar, ten micromolars, or anywhere in between—was a complete mystery. There was no way to estimate how many milligrams of STI-571 a patient would have to swallow to bring the concentration of the drug in the blood to one micromolar or higher. That measurement could be deduced only when the drug entered a human bloodstream.

Almost more important than the predicted minimum drug level, though, was the maximum drug level on which the planning team had agreed. Deciding in advance on a measure at which one can safely conclude a drug doesn't work is common in clinical trial planning; investigators and industry representatives want to know when they can say when. "We are always [asking], 'What are the endpoints that will kill this project?'" said Druker. For STI-571, that endpoint was set at ten micromolars, a far higher concentration than the minimum required for activity. "If we got to ten times our prediction and saw nothing," said Druker, "we would stop the trial."

At each study site, the patients came through the first twenty-four hours unscathed. The relief among the three investigators and at Novartis headquarters was palpable across the hundreds or thousands of miles separating the sites. In Portland, Druker continued to give Romine his daily capsule for the rest of June. Repeated blood tests showed no change in the number of white blood cells or of cells containing the Philadelphia chromosome. Druker knew that, whether quickly or slowly, Romine's white blood cell count would continue to climb. But as much as he wished to give this brave man a higher dose of STI-571, the study protocol forbade it. Plus there was no telling whether the drug would elicit side effects later on. Delayed problems could arise in a few weeks or months. For now, Romine had to stay at 25 milligrams.

A husky man with thick glasses who favored brown suspenders that flanked a respectable but not outlandishly large gut, Romine accepted his fate in the name of helping science. "My feeling always was if it didn't help me, maybe it would help someone in the future," Romine would later write in the *STI Gazette*, a newsletter started by Druker's phase I patients. When Romine and the other first-month patients made it to the end of June with no side effects, the investigators knew it was safe to start the next dose level. But as Romine's white blood cell counts began to escalate, he had to be taken out of the study. It was hard to not see the irony in Romine's letter to Druker years earlier asking to be the first person Druker would call if the compound he'd read about in the *Oregonian* ever went into clinical trials. Enrolling as patient 1 had almost guaranteed that he wouldn't benefit from the drug. His time on the phase I study having reached its abrupt end, Romine started treatment with hydroxyurea, or Hydrea, a chemotherapy drug that was the only other option for patients who had stopped responding to interferon. This drug might keep his counts in check for a little while, but it was just a temporary fix. Eventually it would stop working.

In July, a new round of patients entered the trial at the 50-milligram dose level. The outcomes were the same: no side effects but also no effect on the cancer. Blood samples from these first few groups of patients were revealing in other ways, though. Looking at the concentration of the drug in the bloodstream, Druker could now see that the dose would need to reach about 200 milligrams before the first response would be seen. But if they didn't see any sign of activity at 200 milligrams, he would start to worry. No activity at 200 milligrams meant that the drug probably wasn't going to work.

Druker was in urgent need of a nurse who would help him care for patients and record the profusion of data that were about to be generated. It was only the first month of the study, but so much was needed to prepare for each subsequent month. Patients needed to be walked through clinical trial disclosure forms. Among those who had stopped responding to interferon, blood had to be drawn to ensure the white counts were climbing; the protocol required that they be at least 20,000 and rising for the patient to be eligible for the study because that was the only way to see if the drug was working. Medical records needed to be obtained, and that task alone could require hours on the phone.

Carolyn Blasdel was working as a clinical trial nurse in a Texas cancer center when she decided to relocate to Portland in the summer of 1998. Working on Druker's trial was a perfect match for her. She had the skills, and the work was inspiring. She'd seen listless patients dying of leukemia, seen how they bled out, how their bodies would shut down from the absence of platelets, how they were overwhelmed by infections after their immune system had been shredded by chemotherapy. She knew about the challenges with interferon, and she knew that patients who took a break from it to recover often refused to resume. "I had several people tell me they'd rather be dead than go back on interferon," Blasdel recalled. She knew that CML was an inevitably fatal disease except for the rare few who were cured by a bone marrow transplant. If someone was working on a medicine that could change this sorry state, she was game to help.

Druker met Blasdel on a Saturday when she was house hunting in Portland. They talked in the OHSU hospital lobby for an hour, and he offered her the job. She'd be splitting her time between his trial and another one with a different investigator. She knew her job with Druker would probably last about six months or so, since most phase I trials don't succeed. In early August, Blasdel was ready to begin.

Her first patient was patent number 2, at the 50-milligram dose level. As the study coordinator, Blasdel was drawing the blood, sending the samples for testing, and receiving the results. So she was the first one to see any noteworthy changes in the cancer. Right away, she could see the drug was having no effect on the patient's blood counts. A few days later, patient number 3 entered the trial. This time, the dose was increased to 85 milligrams per day.

Like everyone in the study, patient number 3 had not taken any CML medication for a week or so before starting STI-571, and the white blood cell counts were rising steadily. For the first couple of weeks at the 85-milligram dose, the counts continued to climb. Then one day, Blasdel got the patient's blood test results and saw that the counts had not risen. "Could I have made a mistake?" she asked herself. She thought back over every step she'd taken when she'd drawn the patient's blood. Could she have sent the wrong tube of blood? There was no possible way for her to have done something like that. There was no mistake.

She told Druker, who promptly found out that Sawyers and Talpaz had seen the same thing with their 85-milligram patients.

At 140 milligrams, one patient at each site had a decrease in the number of white blood cells. In Druker's patient, the white blood cell counts dropped to fewer than 10,000 per microliter, a normal level.

At 200 milligrams, the drop in white blood cell counts became even more prominent. All of the patients were having a hematologic response. The abnormal white blood cells, the very definition of leukemia, were starting to disappear. Repeated blood tests showed the counts were down.

There was more. Bone marrow samples from the October group of patients—140 milligrams—revealed an even deeper response to the drug: a decreased number of cells containing the Philadelphia chromosome. They were having cytogenetic responses. In other words, the abnormality responsible for the cancer was disappearing.

Changes in blood counts were important because it was these off-kilter numbers that made people feel sick. Too few red blood cells and platelets and too many white blood cells were all dangerous conditions, and as the numbers returned to normal, patients began to feel noticeably better. But a change at the genetic level signified a deeper response, one that bespoke causes, not just symptoms. A decrease in the number of cells containing the Philadelphia chromosome was evidence that STI-571 was truly a targeted therapy. By hitting the haywire kinase, the drug was eradicating the underlying genetic mutation. This drug, rationally designed against a known target, was doing exactly what it had been created to do. No other drug in history had hit cancer at its genetic roots.

Similar breakthroughs were being made elsewhere. In September 1998, the FDA approved Herceptin, a biologic drug aimed against a protein called HER2, which is overexpressed in some breast cancers. The drug prolonged life and was an undeniable advance for breast cancer sufferers and for cancer medicine as a whole. But Herceptin had to be given with chemotherapy, and it worked only for patients with the HER2-positive variety of breast cancer. Plus the survival benefit had a foreseeable endpoint. In clinical trials, patients with advanced breast cancer who'd taken Herceptin plus chemotherapy lived an average of five months longer than those who'd taken chemotherapy alone. STI-571 took the promise of targeted therapy further. This was a single drug given on its own as a once-daily pill targeting a mutation that occurred in nearly every patient with CML. Would it do for this disease what Herceptin had done for breast cancer? Would it do even more? The original purpose of the phase I trial of STI-571 was to prove the principle behind kinase inhibition: Kill the kinase and you kill the cancer. Just months after launching the trial, it was much too soon to think about the broader implications. And yet it was impossible not to.

Druker, Sawyers, and Talpaz held weekly conference calls, joined by John Ford, who was monitoring the trial from Basel. When the counts began dropping at 200 milligrams, even Talpaz, the seasoned investigator who'd witnessed many new agents rapidly elicit promising responses that soon faded, was impressed.

Hesitation lingered, though, as everyone watched for side effects, bracing themselves for disaster. Nearly every single cancer treatment available at that time did something terrible to the body as it killed the malignancy. The theory that a drug aimed against one single kinase would not cause such widespread damage was plausible, but it was hard for the investigators to truly believe it because everything they'd experienced said otherwise.

By a few months into the study, none of the patients had developed any severe side effects. Many had puffiness under the eyes, a result of fluid retention, and in the waiting room patients swapped home remedies to reduce the bagginess, including tea bags and hemorrhoid cream. But the problem was more of a curiosity than a concern. Compared with the side effects of chemotherapy or of interferon, which most of the phase I patients had taken, puffy eyes seemed a small thing.

There were some problems. Patients pushed through dramatic leg pain at the start of therapy as bones swollen from the profusion of immature white blood cells returned to normal. Cramping occurred because trial patients had to take the pill on an empty stomach to prevent food interactions. There was some tiredness. But the debilitating side effects that had made cancer treatment so feared were absent. A patient on chemotherapy was certain to experience some harrowing combination of a number of them: numbness, kidney damage, liver damage, nausea, vomiting, hair loss, peeling skin, rashes, diarrhea, constipation, headaches, changes in blood pressure, dry mouth, insomnia, fever, fatigue, bloody urine or stools, anemia, memory loss, pain, swelling, problems with urination. Drugs could ameliorate some of the problems, but no one taking chemo came through unharmed.

The STI-571 experience wasn't normal. This wasn't how cancer treatment worked. Patients were supposed to be doubled over in agony, vomiting into waste bins at the side of their bed, too weak to take care of themselves. These patients were carrying on with their lives as if nothing was wrong. Chronic flu symptoms and depression had plagued many of them during their stretches on interferon. In light of those dark months, STI-571 was a dream.

Still, Druker barely allowed himself a modicum of excitement. He knew that what he was seeing with the phase I patients could disappear in an instant. STI-571 might be working now, but in another week all those blood counts could escalate right back to where they started. Some late side effect could appear that would render the drug useless. Throughout the ninety-hour workweeks he put in during the study, Druker never stopped to think about the fact that this drug was a success. He couldn't. Not yet.

As if vindicating his caution, one of the patients on the 200-milligram dose developed liver damage. Some people at the company panicked. They wanted the investigators to go back to 140 milligrams. The investigators vehemently disagreed. "Don't go back, enroll more," they insisted. The only way to know whether the toxicity was a one-off situation or a real problem was to continue giving the drug to people at the current dose. They had already seen some activity with the drug, and the only thing to do was keep moving forward. Novartis agreed to continue at 200 milligrams, allowing the investigators to enroll two patients per month starting in January 1999 instead of raising the dose until the problem was resolved.

26

THE ONE THING THEY DIDN'T HAVE

The investigators' insistence on moving forward with the study was quickly legitimized. The ailing patient's liver problems resolved and disappeared, and the patient continued as a participant. The dose increases resumed. No other patients came down with liver problems. Druker decided the time had come to enroll patients who were considered slightly more difficult to treat. The study was still restricted to chronic-stage patients, but now he wanted to see how the compound would work on those who had trickier white blood cell counts or other complications. Judy Orem fit the bill.

Orem, who grew up in Portland but had settled down with her husband Frank and raised two children in California, was 35 when her grandmother, May Belle Roscoe, died of leukemia in late 1979. Roscoe had been in and out of the hospital as her chemotherapy grew less effective. The previous April, she began hallucinating and demanded that Orem remove the flowers she'd brought her, believing them poisonous. She thought the nurses on the floor were conspiring against her. In a moment of lucidity, she decided to stop all medication. Her doctor warned her that she'd have about three days to live, that her blood counts would double, then triple, then quadruple, and she would die from an aneurism, which was exactly what happened. Her death certificate reported the cause as CML.

In 1990, Orem's mother, May Belle's daughter, was diagnosed with lymphoma, which upon a bone marrow biopsy turned out to be chronic lymphocytic leukemia, or CLL. Like CML, CLL begins in the bone marrow, but the cancer grows in the lymph cells, a different type of white blood cell than is affected in CML. Leukemia isn't inherited. There are no genes passed down from one generation to the next that carry a leukemia trait. The Philadelphia chromosome is spontaneous, a so-called somatic mutation impossible to predict. Having a mother or grandmother with leukemia, or even both, has no bearing on the health of the daughter or granddaughter. Radiation exposure can increase risk, and the disease occurs more often in men than women and more often in people 65 and older. But a family history of CML has absolutely no bearing on whether or not an individual will develop the disease.

Still, a few years after her mother's diagnosis Orem started worrying about her own health. The coincidence of having her mother and grandmother get blood cancers left her with the lurking fear that she would be next. In December 1995, during a routine physical, she asked her doctor to do a blood test. The next day, Orem's doctor called to tell her she needed to see a specialist right away. Her white blood cell count was 66,000, more than 50,000 cells above the upper limit of the normal range. The news was devastating. The fact that she had been almost expecting this to happen to her did nothing to mollify the shock. Her arms and legs grew numb as she hung up the phone. She had leukemia. She was going to die from cancer.

"No, this is wrong," her husband said when Orem told him. "You're supposed to bury me, not me you." They had met during intermission at a local theater's Gilbert and Sullivan production when she was in high school and he was in college. Forty years later, he could not accept the thought of her dying first.

A bone marrow biopsy immediately confirmed her diagnosis. Orem had CML. The disease was still in chronic stage, with only 5 percent blast cells, the immature, malfunctioning white blood cells that increase as the disease progresses to the blast crisis stage. After Christmas passed, she started on interferon, giving herself shots at home each day, each syringe filled with the standard 3 million units of drug, an amount that could be increased if the drug's benefit began to wane. She continued at her job managing the office at a program for vision-impaired students while she got the hang of managing her medicine, increasing from 3 million units to 6 million and eventually to 9 million. She joined a support group that her minister had recommended, finding a fellow CML patient who, then five years after her diagnosis, was able to reassure Orem that she would be fine. "That was the most important thing I ever heard," Orem said.

Orem decided fairly early on that she did not want a bone marrow transplant, knowing the procedure could so easily go wrong. "I would rather be good for two or three years and face death than go through that," she told her husband.

Over the next three years, Orem continued the interferon. The drug, which works by stimulating the immune system, had proved incredibly effective at lowering white blood cell counts in people with CML for a certain amount of time. But almost always, patients with CML would develop resistance, and the white counts would start rising. Or the side effects of the drug would prove too debilitating. Interferon, a synthetic drug derived from an immune system product, triggers an immune response in much the way that its naturally occurring counterpart activates our immune systems to fight a viral infection. The similarity shows in the side effects: When people take interferon, they often develop a fever, chills, and a paralyzing fatigue, like having a flu that lasts for months or even years at a time. The drug can also cause severe depression.

Orem lost weight and was often tired while she was taking interferon. She had such a strong aversion to smells that walking through a grocery store became torturous. Her short-term memory became uncharacteristically faulty; if she paused mid-sentence, she'd have a hard time picking up where she left off. As it was for so many CML patients, interferon became the defining hallmark of the disease. Many times, family members wouldn't remember details about the cancer, but they would remember the decimation wrought by the drug: the tiredness, the persistent flu-like symptoms, the depression. But for Orem, as for so many CML patients, the drug was working well enough to continue, keeping her white blood cell counts at a safe level.

In 1997, Orem's friend, who worked as a medical technologist in Portland, had heard a radio item about a doctor up the hill at OHSU who was studying a new drug for CML. Druker had received funding from the Leukemia and Lymphoma Society, a fact that bolstered the friend's interest and confidence in his work. She had taken on the role of surrogate nurse for Orem and promptly called Druker's office to give them Orem's phone number. Orem and Druker talked, and he put her on his list of possible trial candidates. But, he explained to her, she wouldn't be eligible for the study unless the interferon stopped working.

In June 1998, just as Bud Romine was taking the first human dose of STI-571, Orem's medication stopped working. Her white blood cell counts were rising. Her doctor added a chemotherapy drug called ara-C to boost the power of interferon, but it didn't work. Druker, who'd been keeping tabs on Orem's condition, asked her to make the trip to Portland for a preliminary examination. She was eligible now, but her white blood cell counts had to reach 20,000 before she could enroll.

Until now, Druker had restricted the trial to those patients with slowly rising counts. He knew that if the drug didn't work on someone whose blood counts were rising rapidly, then the patient would die very soon, and their enrollment would have been in vain. Plus, people with slowly rising blood counts are easier to manage when side effects arise. But with early indications that STI-571 was working, Druker felt bold. For him, that didn't mean disclosing the news or letting himself believe what was happening; it meant enrolling more challenging cases.

That September, Druker told Orem she could enroll in January. He had no idea whether the drug would work for her. Orem's doctor wasn't crazy about the idea. "Let somebody else try it first," she'd counseled. But Orem knew she didn't have that long to wait. Frank encouraged her to go for it. He knew that going back on interferon was not an option, and he felt confident that the kind doctor up in Portland would stop the treatment if it was harming his wife. Believing she had about six months to live, she and Frank took their children and grandchildren on a trip to New Zealand. "It was a family memory trip," said Orem. It was the kind of thing they had talked about doing when they retired.

After a close call with a dangerously high platelet count, Orem returned from New Zealand with blood cell counts that made her eligible to start the trial in January 1999. Now allowed to enroll two patients each month, Druker also began treating a woman whose white blood cells counts had escalated from 20,000 to 125,000 in a single week. These two patients raised the stakes. Treating asymptomatic, easily managed patients with STI-571 was interesting enough. Could the drug work in a less stable situation?

Within three weeks, both patients' blood counts had returned to normal. Orem was astonished. After all, it was only by chance that her friend had heard about the trial, and she had come very close to staring death in the face. By February 1999, just weeks after entering the trial, she was flush with energy. "At that point, I knew we had something that nobody had ever seen before," said Druker. "It was just absolutely amazing."

Still, it wasn't enough to leave Druker feeling confident about the drug's efficacy. Day after day, he looked after his patients. For months, they were all he thought about. He spent hours explaining the drug to them, listening to their concerns, responding as best he could, and opening his office to them for visits. But kept hidden behind his reassuring bedside manner was the persistent, gnawing question: *Will this last?* The question weighed on his mind nonstop, and he would not allow himself for a second to think the answer might be yes.

The hematologic responses continued. The cytogenetic responses—decreases in the number of cells expressing the Philadelphia chromosome—were also happening, though less rapidly. The effects were being sustained. The side effects followed the same consistent pattern, with nothing that was really intolerable. There was the initial stomach cramping that subsided, the pain as swollen bones settled back to normal, and the water retention that showed up in puffy eyes. The first few days on the drug were often difficult, but there were no persistent side effects requiring medical attention. It was the polar opposite of chemotherapy. "The problem that I was having was it was too good to be true," said Druker. "If I told you I could give you a pill that was going to put your cancer in remission with no side effects, would you believe it?"

In April 1999, the protocol was amended again. The earliest trial patients, those who had enrolled at dose levels too low to help, were allowed back into the study at the effective dose of 300 milligrams. Bud Romine, patient number one, returned to OHSU. The minister from Bakersfield returned to Sawyers's clinic. The drug worked in both of them.

A few weeks after Romine's return, Druker was holding his usual clinic hours. From the start, he'd set aside time to meet with each phase I patient one-on-one. He wanted to make sure they had a chance to ask questions about the drug and to have their questions answered. One day in April, Druker opened his office door still carrying the knot in his stomach that the drug's benefit might come to an end. He was dreading that moment, waiting to wake up from what had to be a dream. "I was trying to protect myself from getting too excited," he said.

Bud Romine sat in a chair across from Druker. He was responding well to the higher dose of the drug. His blood counts were normalizing, and there were signs of a cytogenetic response, too. When Druker asked Romine how he was doing, the words tumbled out. The sword that had been hanging over his neck, waiting to fall, was now gone, Romine told him. He'd been waiting to die, and now he knew he was going to live. The next patient was Judy Orem. She talked to Druker about how she had resumed making plans for the future, her hope now restored. Another patient talked about how the black cloud over his head had been lifted. The drug had given him his life back.

A resistance inside Druker suddenly melted. For the first time, he allowed himself to believe in what was happening. As he watched the tears roll down their cheeks, Druker shed his own as he finally allowed himself to believe what was happening. "I realized they were so far ahead of me," said Druker. "They already accepted that this drug had worked and had changed their lives."

Before starting the trial, these patients had been told that there was nothing more that could be done, and that they should say their goodbyes and tie up the loose ends of their lives because their time was limited. For Druker, that day in his clinic was a first glimpse at what this drug had really done: "It was to give these patients the one thing they didn't have, and that was hope."

For some of the phase I patients, their restored future came with a strange twist. Having been told they were going to die, some had sold everything, packed their lives up so as to leave nothing behind for anyone to deal with. They'd quit their jobs and spent their savings. Now, with life returning to normal, they were finding themselves at an unexpected loss. Still, no one complained. It was enough to be alive.

EARLIER IN 1999, Druker and the other investigators had flown to New Jersey to meet with Novartis executives at the company's US headquarters. They wanted to expand the study to include blast crisis patients, the ones who had only weeks or months to live. In a typical phase I study design, these would have been the only patients enrolled. Novartis's decision to focus on early-stage patients, though motivated by the company's wish to get the fastest possible reading on the drug, had turned out to be ideal. But Druker and the others felt that if STI-571 could save lives, they needed to get it to the sickest patients. "None of us thought the drug would work in blast crisis," said Sawyers, "but if it did, we needed to know." Novartis agreed.

These late-stage patients began enrolling in the study in the spring of 1999. At OHSU, UCLA, and MD Anderson, patients confined to wheelchairs and attached to oxygen tanks were brought in by family members, all of them hoping for something that would help. No one was expecting a miracle, maybe just an extra month or two.

The responses began within a week after starting STI-571. Among several dying patients, white blood cell counts dropped, making room for restorative red blood cells to proliferate and heal the body. The color returned to their faces. They gained strength. They got up out of their wheelchairs and walked out of the hospital.

The results were mind-boggling, but they also set off alarm bells: This drug had to reach as many CML patients as possible as quickly as possible. The only way to do that was through a phase II clinical trial. The next step in development, the phase II study would enroll more patients at more sites, giving more people access to the drug.

Phase I had room for dozens of patients at most; at phase II, hundreds of patients could enroll. The company could have a phase II study for chronic-stage patients and a phase II study for those with more advanced disease. And it was the next necessary step toward submitting the drug for FDA approval, which would allow for the widest release of the drug. The company would need at least a year of phase II data for the regulatory review, so there was no time to lose. In a normal clinical trial, phase II doesn't begin until phase I ends. There is time to reflect on the data collected from those first patients, to decide on the best dose and best approach for phase II. But here, that approach made no sense. "Every dose was working after we got to 350 to 400 milligrams," said Sawyers. "Why do we have to keep going? We've got to stop and do the phase II."

In June 1999, as Sawyers, Talpaz, and Druker were on their way to Basel to begin planning the phase II study with Novartis executives, Sawyers received the results from one of his chronic-stage patient's latest genetic tests. He stared at the numbers on the page: 0 out of 20 cells containing the Philadelphia chromosome. It was the first complete cytogenetic remission, a landmark moment for the trial. All of the cells with the cancer-driving genetic mutation had been eradicated. It was what everyone—the doctors, the patients, the company—had been hoping for. A lowering of blood counts was fantastic and a sign of improving health. A disappearance of the mutant gene was the promised land. Unable to reach Druker, who'd already boarded his flight, and unable to contain his excitement, Sawyers called the patient. "We knew we had a home-run drug at that point," he said.

There were some disappointments. One or two early-stage patients did not respond to the drug, their blood counts failing to be reined in. The most dramatic failures were among the blast crisis patients who had been brought back from the brink of death. After the sudden recoveries, many of them relapsed just as quickly. "The patient who was in the wheelchair, on oxygen, [and then] walking and dancing . . . two weeks later, [was] back in the wheelchair and on oxygen," said Sawyers. The swiftness and severity of the relapse shocked Sawyers, especially in the wake of the elation over how well the drug was working.

Sawyers wanted to understand what was going wrong with the late-stage patients. Why would they respond so well at first and then relapse? What was different between those patients and the chronic-stage patients? "It was absolutely essential to understand the molecular basis for that," he said. So, as he continued to run the phase I and then phase II studies, he also returned to the lab. Sawyers wondered if understanding the late-stage patients might also shed light on a rare few chronic-stage patients who had the Philadelphia chromosome but didn't respond to STI-571 at all.

THE LOSSES, MAINLY concentrated in the blast crisis patients, were uncommon. For the most part, the positive responses were sustained. If the drug kept working, then people with CML could live normal, healthy lives simply by swallowing a tiny pill once a day.

Nobody used the word "cure," though. The measurements charted in the phase I study were not proof that the drug would extend lives. The true clinical benefit of the drug could be gauged only by whether, and by how much, it prolonged life or reduced symptoms. Hematologic and cytogenetic responses were surrogate end points, stand-ins showing that the drug was doing things that should, in theory, extend survival and reduce suffering. Considering that it was the profusion of white blood cells that ultimately led to a CML patient's demise, measuring those numbers along with the number of cells harboring the genetic mutation known to cause the disease seemed fairly reliable indicators of the drug's benefit. But that wasn't a cure. If a patient stopped taking the drug, the disease returned. More to the point, not enough time had elapsed to see if STI-571 helped CML patients live longer, healthier lives than they would have without the drug.

Still, for the patients and clinicians, the outcomes had already far surpassed all reasonable hopes. Blood counts were being restored. The bad gene was disappearing. Patients were feeling better. It was time to open the phase II trial and to tell the world about this drug.

27

BUZZ IN THE CHAT ROOMS

The phase I patients were increasingly aware that they were part of something special. In Oregon, they became friends and, in the summer of 1999, started a monthly gathering. They'd share a meal and hash over the details of their illness, how they had arrived at Druker's doorstep, how they were feeling. They spoke a secret language, where words like "FISH" and "blood counts" carried the weight of their survival. They felt charged with responsibility, as if knowing the science were somehow their job now. They were ordinary people who'd been diagnosed with a blood cancer and had found themselves in a rare circumstance: the first initiates in an entirely new kind of medicine.

The chatter migrated online. In the late 1990s, Internet chat rooms were in their fledgling years. This kind of forum was perfect for the CML community, allowing its members to exchange information about the disease, about its treatment, about how to handle the various side effects. More experienced patients enjoyed being able to pass on their hard-earned wisdom. For many people, local cancer support groups included people battling all types of malignancies. There were rarely enough CML patients in a single county to form a local support group. Online, however, the conversation was confined to their disease only, and so the talk proved much more appealing and useful. People got to know one another through message-board postings, and they opened up about the emotional challenges of their disease and the rest of their lives, too. They were a support group of people who each knew exactly what the others were going through.

A message board started at Egroups.com by a man named Robert Neil, whose mother had been diagnosed with CML, was the earliest of the many incarnations of online CML forums. There were a few dozen active members, and many silent readers who soaked up the information. Among the group were a few people who were enrolled in the phase I trial. Naturally, they posted their test results and recounted conversations they'd had with their doctors. They talked about how they were feeling. Nearly everything they wrote was positive.

The message boards were open to anyone. The trial investigators, who were aware of the forums, suddenly found themselves in the middle of a new phenomenon: Internet sharing.

Each day, the investigators would return home after a long stretch at the clinic and find confidential conversations recounted verbatim online. The information gave hope to people with CML who were struggling with chemotherapy and interferon and had not known about the trial or been able to reach a study site. Because of the World Wide Web, many patients with CML now knew more than their doctors about this experimental new drug.

The positive reviews stirred up demand among those not enrolled. It wasn't long before Druker, Sawyers, and Talpaz were hearing from hundreds of people from around the world who'd been diagnosed with CML. They'd read about the trial on the Internet, seen the incredible data, and they wanted in.

Having had minimal, if any, exposure to web-educated patients, the investigators had no plan in place for how to respond. Should they water down their messages in the clinic, knowing that whatever they said to their patients was going to be broadcast far and wide? That might be wise. But the drug was working. Why shouldn't everyone know about it? Plus, censoring the conversations in the clinic wasn't their style. All along, Druker had made himself available for any questions, and that openness was inherent to the intimate nature that his phase I group had developed.

They also wondered if the publicity might serve another purpose. It was time to ready everyone for the next clinical trial, and Novartis was not moving quickly enough. Maybe the freewheeling chatter about this new drug that was melting cancer away would give the company a kick. "It was important . . . for Novartis to feel the heat, to get their shit together, and start really ramping up the project," said Sawyers.

The plan, as Sawyers knew, was to open three separate phase II studies: one for chronic stage, one for accelerated stage, and one for blast stage. As with the phase I study, these trials were restricted to patients who had already tried interferon, the medication either proving intolerable or ceasing to work. Cancer drugs are rarely approved for all stages of cancer at once; rather, new medicines are approved for each specific population in which the drug was studied. When the FDA first approved the drug—if the FDA approved it—the first indication would be for patients for whom the current best treatment option failed. These three phase II trials, which would enroll a thousand patients altogether, would provide the evidence for those approvals. After that would come phase III, a large and lengthy study in which newly diagnosed patients would be randomly assigned to treatment with either STI-571 or interferon plus ara-C. Only at phase III would the drug be tested on patients who'd never been on any other CML medication.

But there was a problem. According to Novartis executives, there wasn't yet enough of the drug to start a phase II study. And there was no telling when there would be enough. Sawyers wondered if the message board chatter might stir the publicity-fearing pot at the corporation.

Every day, there were new posts from the members lucky enough to be in the phase I trial. Every day, there were calls to OHSU, UCLA, and MD Anderson from patients interested in enrolling. And every day, there was nothing new to tell them. The trials were full. There was no more drug. They would have to wait.

28

SAVING THEIR OWN LIVES

Because the transition from the phase I study to the phase II study was so quick, Novartis had not had time to produce enough STI-571 by the time the investigators wanted to start enrollment. A few kilograms of STI-571 had been manufactured for the phase I study. Hundreds would be needed just to launch phase II. The clinicians and the industry representatives had talked about the phase II study since February 1999, but it wasn't until the summer that the company realized the shortage it faced. "They hadn't planned for success, they had planned for failure," said Druker. That mind-set had left the company unprepared for the next steps.

It wasn't just that the phase I study had been designed to give a quick readout about whether the drug worked; many people had assumed it wouldn't work. The assumption was understandable. Fewer than 10 percent of experimental compounds tested in phase I studies are ultimately approved by the FDA. This compound had killed dogs in the toxicity studies. Targeting a kinase was an untested strategy. While some people at the company pushed to move ahead, others dampened the enthusiasm by pointing again to the reality of the market. The sales projections were dim. Betting on failure was much more reasonable than betting on success. By the summer of 1999, those miscalculations had caught up with the company and were delaying the delivery of a breakthrough treatment to patients who needed it.

At a meeting in Bordeaux, France, in the summer of 1999, Druker discussed some of the data from the phase I test so far. John Goldman, the British leukemia expert who'd been excluded from the first trial, was in the audience. Stunned by what he heard, Goldman asked John Ford when he could get the compound to start a trial with his patients. "Well," Goldman later recalled being told, "Novartis isn't planning to develop it. Maybe in a couple of years." Goldman wrote a letter to Daniel Vasella, then the CEO of Novartis, urging him to move forward, but he never heard back.

Druker was doing all he could to speed up the process. His only contact at Novartis was Ford, who was bound by what the company was telling him.

"Who can we talk to, to try to accelerate this?" Druker asked him.

"Me," said Ford.

"Who do you work for? Can we talk to them?" Druker pushed back.

"No," he recalled Ford replying.

"But you're not fixing this drug supply shortage."

There was nothing Druker could do. Part of him understood. "Coming into this new way of treating cancer, a new category of drug, a small market projection, who in their right mind would put a lot of time and effort into that project?" he said later. But with the phase I data now in hand, it was time to change that thinking. "We were moving at warp speed, and we needed a champion inside Novartis."

Druker was entirely frustrated. There were no more phone calls he could make. There was no one he could talk to who could influence Novartis to get moving. All he could do was stare at the growing list of patients who wanted the drug.

And run. He joined marathons. He ran to work, up the steep hill to OHSU. After long hours at the clinic, he ran home. It was as if the speed he longed to see in the clinic was demanding another outlet. So he ran.

Vasella tells the tale differently. "There was no drug shortage," he said. And the situation was not unusual for an experimental drug. "You never have that," Vasella said of the need for an ample quantity at the very start of phase II. The only difference with STI-571 was that the need had arisen much sooner than usual, leaving the company with an "unexpectedly high demand." The phase II trial was being set in motion unusually fast, before the phase I had even been completed. That timetable was unforeseen and hadn't left the company with enough time to make the necessary amount of the drug.

Vasella acknowledges the continued resistance within the company, but also how important it was to him to take the risk. Jörg Reinhardt, who was the head of the development team at the time, was insisting on proceeding with caution, telling Vasella about not having this or that bit of data or the manufacturing capacity. But having seen the phase I results, Vasella wanted to plow on. "Come on now," he told Reinhardt, "move ahead." When Reinhardt expressed concern about the cost, Vasella dismissed it as unimportant. "The problem is somebody has to be willing to take a risk and accept that he might get the blame for failing," Vasella said later. As he tells it, after his initial intervention that moved the drug from the preclinical phase into clinical trials, Vasella continued to be its champion. Though uninvolved in managing the details of what it took to manufacture sufficient amounts of the drug, his willingness to take the risk of developing STI-571 provided cover to those upon whom those responsibilities fell.

In August 1999, the phase II trials for accelerated and blast crisis patients opened. Novartis had managed to produce enough drug for that population. It would have been considered highly unethical to withhold a drug from those patients if it clearly could be made in some quantity. Patients with accelerated or blast crisis CML could enroll in the phase II trials without restriction.

But the next trial for chronic-stage disease remained on hold. Patients in this early phase of the disease worried that the leukemia would progress before they began treatment. Even though drug access for later-stage patients was guaranteed, no one wanted to wait until they got to that stage because it might be too late. They wanted the medication when it could do the most good. "Their view," Druker said by e-mail, "[was] that Novartis had essentially crafted a death-bed strategy." That the company had launched phase II studies for blast crisis and accelerated disease was evidence of moving forward with development, but chronic-stage patients stuck on interferon wanted what they knew was a superior treatment as soon as possible.

ONE OF THE patients on Druker's list of those wanting access to STI-571 was Suzan McNamara. In March 1998, when she was 31 years old, she began experiencing bone pain and rapidly lost 15 pounds. A blood test revealed her white blood cell count to be 380,000. She had CML. Petrified by the prospect of a bone marrow transplant, she opted for interferon. The treatment worked for a while, lowering her white blood cell count, but eventually she had to quit her job because she felt too sick to work. Each day, she injected herself with interferon and ara-C, and each day she felt worse. She was depressed, she couldn't sleep, and her hair was coming out. She could barely eat, and she continued losing weight.

McNamara had stumbled across the Egroup.com forum during an online search for information about CML. She read the posts by the few patients on the phase I trial for STI-571. Their test results were encouraging, and the patients reported that they felt good. McNamara, whose life by then revolved around her bed and the computer two feet away, seized on those words.

The trial patients often talked about their progress in meticulous detail. People on the forum who'd attended meetings at which the investigators discussed the drug reported back what they'd heard. Everyone spoke positively. McNamara asked her doctors about enrolling in the trial, but they discouraged her. A million drugs went through phase I tests, they told her. Why would this one be any different? Still, she could not ignore what she was reading. All of the non-trial CML patients reading those posts wanted in, and they all knew they had to wait until they relapsed. Some got creative, finding ways to expedite the process—say, insisting that they could not stand interferon's side effects anymore—even before their white blood cell counts began their inevitable ascent. They knew the criteria that would make them eligible for the trial, so they made themselves eligible.

For McNamara, relapse was all too real. In October 1999, her treatment stopped working. About six months after starting treatment, her platelet level fell dangerously low. A lower dose of interferon and ara-C didn't work. She knew that it was only a matter of time before she entered the accelerated stage of CML. Now approaching 33, she could be dead in a year.

She took printouts of the online reports from the phase I trial patients to her doctors. Finally, they agreed to call. There was no good news. The trial was full and closed to further enrollment.

Out of sheer desperation, McNamara sent an e-mail to Druker telling him her story. He replied within eight hours. He was sympathetic, but he told her there was not enough drug. It would be four or five months before she could enroll in the next trial, but Druker wasn't even certain that would happen. "We are fearing it might close down eventually," McNamara recalled him as saying. "Maybe as a patient you can do something."

The moment sent a shaft of light down to McNamara, stuck in the depths of her illness. His words "lit a fire in me," she said. "There is no way that there is a drug out there that can help me that's not going to be given to me."

She lay in her bed trying to think of what she could do, something big that had a real shot at working. She remembered something she'd seen on the Internet and wondered if it might work: an online petition.

At the time, the idea was still fairly novel. Although accessing the web from home was not uncommon in the late 1990s, many of the innovations that people now take for granted had not yet evolved. Not every advertisement included URLs. Not everything could be found online, and websites certainly had not become the vehicle for patient advocacy that they are today. The Internet had not taken off as a social medium, and its potential power to generate social change was unforeseen. Yes, the trial patients were posting all the minutiae of their experiences with STI-571, disclosing intimate details about their health and personal lives. But their readers were not mobilizing online to get themselves the drug. Forging a new kind of online alliance wasn't their intention. Although most of them had witnessed the AIDS crisis during the 1980s and seen the power that patients could have when they banded together, this cancer was different. It was a rare disease, not an epidemic, and there was no social charge behind the cause. They weren't angry. They just wanted the drug.

The online CML community readily embraced McNamara's idea. The petition, which called for the manufacture of STI-571 to be sped up, was posted online on September 21, 1999. Within three weeks, 3,300 people had signed it, and the signatures continued to amass. Patients, caregivers, friends, and friends of caregivers had all logged on to this call to Novartis to make more STI-571 and get it to the patients who needed it. Members of the Egroup message board wrote a letter to accompany the petition and addressed it to Daniel Vasella, then the chairman and CEO of Novartis. The letter read:

> Through extensive interaction on the internet, a large number of CML (Chronic Myelogenous Leukemia) patients are well-informed about the results of the Phase I and Phase II trials of the new drug STI-571 which is produced by Novartis. The sources of information are varied and include published papers on the drug, papers given at professional meetings, and the direct experience and knowledge of a significant number of patients who are taking part in the trials. . . .
>
> As you know, the drug has shown no significant toxicity to date, and the results, in terms of both hematological responses and cytogenetic responses, have been more than impressive at this early stage. . . . There is also an acute awareness that the drug is in clinical trials and a realization that there is a need to be cautious at this relatively early stage. . . .
>
> Nevertheless, many of us who have signed the petition believe that the theory behind this new drug is fundamentally sound. . . . It is not impossible, based on results to date, that the new drug will prove to be a functional cure for some patients. . . .
>
> Because of the particularly good prospects for this new drug, we have viewed with growing concern our belief (based on information from various sources) that the supply of the drug has not been sufficient to expand the trials as fast as the evidence to date would warrant. . . . There are many CML patients undergoing difficult treatments, in some cases with considerable suffering, and where the outcomes are not at all assured. . . .

> We therefore ask for your assurance that everything will be done to produce a sufficient supply of STI-571 to ensure that the trial investigators are not held up in any way at all in trialing this new drug, and in advancing to the certification that we anticipate.

In October 1999, it was time to send the message to Novartis. Meanwhile, Druker was preparing for that year's ASH meeting, at which he would present the results thus far from the phase I study. The first few months' worth of data had been presented at the 1998 ASH meeting, but now the trial had been going on for more than a year. When the investigators submitted the study for consideration back in August, they had six months of data at the 300-milligram dose level. The responses were so impressive that the ASH advisory board invited Druker to present the study at the meeting's plenary session. It was the highest honor the organization could give. Four years earlier, Druker had presented the preclinical study to an audience of about fifty, hoping someone would take interest. Now he would be speaking to an audience of around 20,000 clinicians from around the world. Every major news outlet was going to cover his report.

The CML patients clamoring for access to STI-571 also knew about the meeting. They knew it would be the perfect place to call out Novartis on how long it was taking to make this lifesaving drug more widely available. If they put out a press release during the ASH meeting, Novartis would come under harsh scrutiny and would have no choice but to provide the drug as fast as possible. A petition sent only to Novartis, without a press release, might not raise the stakes enough. They asked Druker, whom they loved and to whom they would always defer, to weigh in.

Druker was conflicted. He knew that if the petition was made public, the message surrounding STI-571 at the ASH meeting would be about the drug shortage. But this meeting would be the first time that any results from the drug trial would officially be made public. Shouldn't the message be about this unprecedented breakthrough? He advised the patients against the press release.

His counsel was also strategic. The petition gave him some much-needed leverage with the company. He alerted the executives about the letter they'd be receiving from patients. He also emphasized that he needed an announcement about the phase II trial from Novartis. "If I'm going to give a presentation, there has to be a clinical trial these patients can get into," he explained. He turned their attention to the media nightmare that would ensue if a trial were not scheduled.

Novartis got the message. On November 2, 1999—McNamara's thirty-third birthday—she received a call from Druker. "Suzan, I just got off the phone with Novartis," he told her. "Your petition did a world of good." The company had agreed to speed up production and open a trial in the next month or so. A week later, she received a letter from James Shannon, a representative for the company, confirming the start of the next study. "An international multicenter phase II study is planned that will open for enrollment in January 2000, if not sooner," Shannon stated. "Currently limited by availability of drug supply, Novartis has devoted substantial attention to making sufficient quantities of the agent available as soon as possible." People with CML had rescued a drug that would, they hoped, rescue them.

According to Vasella, the company had already been ramping up production of STI-571 for the phase II trial of chronic CML. There was no way the company would not move ahead with a drug that was potentially valuable. "Would you stand in front [of people] and say, 'We have a potentially lifesaving drug, and I'm not going forward?'" he asked, the notion obviously ridiculous. He knew there were patients who might benefit from the drug. He also knew that in the business of developing new drugs, "more are failures than successes," he said; risk of failure is part and parcel with working in the pharmaceutical industry.

As Vasella told the story some years later, the company had always planned to produce STI-571. McNamara's petition had only given the company a nudge to speed things up a bit. The availability lagged only because the trials were moving so quickly, not because the company was moving slowly or shelving STI-571 for good.

"Dr. Druker will stand by the fact that it was file 13, never to be seen again," said McNamara. "For me, I'll never know."

29

A RESPONSE RATE OF ONE HUNDRED PERCENT

A few weeks later, Druker stood on a podium at the ASH plenary session. Having seen the preliminary data a year earlier and caught wind of the results that Druker would now present, the news media had demanded that ASH release the data from its embargo so as to allow newspapers and television to report on the study on a Friday, two days before Druker's presentation. By the time that session came, the drug was already making headlines in major US newspapers, and all the participants were talking about tyrosine kinase inhibition. Under the spotlights of the New Orleans Convention Center's cavernous hall, row upon row of chairs filled by cancer and blood doctors and researchers, Druker presented the slides he'd prepared for the occasion, his midwestern drawl adding to his understated manner. He didn't need to say much; the numbers spoke for themselves.

Druker took the audience through the study. All of the enrolled patients were in the chronic stage of CML, with fewer than 15 percent blast cells. All had stopped responding to interferon, which meant they'd had no significant change in their blood after three months of treatment, had no changes in the number of cells containing the Philadelphia chromosome after a year, had lost whatever response they'd been having, or were intolerant. All of the enrolled patients had ceased treatment for at least one week before starting STI-571.

The trial, reported Druker, was a standard dose-escalation study. Eleven dose levels had been tried, starting at 25 milligrams and reaching 600 milligrams by the time of the ASH meeting. The average age of the sixty-one patients who had entered the study was 57 years. They'd been taking STI-571 for an average of 190 days. The duration of their disease was variable. Some had received the diagnosis less than a year before entering the study. At least one had had CML for more than thirteen years.

Druker addressed the toxicities first. "No dose-limiting toxicity has been encountered," his slides read. A few patients had mild myelosuppression, a decrease in activity inside the bone marrow that resulted in diminished production of red and white blood cells. Forty percent had experienced low-grade nausea, a problem that faded as the body grew accustomed to the foreign substance. There had been a bit of muscle cramping and a bit of swelling—mostly puffy eyes—but that was it, he told the crowd.

Druker pressed the button on the remote, and the slide projected on the screen behind him changed. Hematologic Responses. The chart showed how many such responses had occurred at each dose level. At the 25- and 50-milligram levels, no changes had been observed. At 85 milligrams, one out of four patients had a response. At 140 milligrams and higher, 100 percent of the patients—all of the fifty-one patients who had come through the entire study so far—had experienced a hematologic response. The audience gasped.

Druker pressed the button again. Complete Hematologic Responses. Everyone's eyes floated to the bottom of the giant screen behind him. Among the patients who'd received doses of 300 milligrams or higher, every single one had had a complete hematologic response. Their blood had reverted to a completely normal state.

Graphs charting the changes in white blood cell counts showed how the levels flatlined as the study progressed, plummeting from 20,000 or higher to well below 10,000. One patient whose counts had been nearly 100,000 at the start of treatment was now at about 6,000. The significance was obvious. Everyone in the audience knew that anything below 10,000 was considered normal.

He pressed the remote again: Cytogenetic Responses. At five months, 45 percent of patients—nine out of twenty receiving 300 milligrams or higher—had a cytogenetic response, ranging from minor to major to complete. The number of white blood cells with the Philadelphia chromosome had vastly diminished, in some cases disappearing entirely.

Druker showed a slide summarizing the data, and a last one to acknowledge many of the people behind the work, those who'd brought the drug to this very moment: Elisabeth Buchdunger, Jürg Zimmermann, Alex Matter, Nick Lydon, the members of his lab at OHSU, Grover Bagby, John Goldman, Charles Sawyers, Moshe Talpaz, Tom Roberts, and Jim Griffin. Many of them were among the transfixed crowd.

Druker thanked the audience for listening, and the talk ended. The room erupted in applause.

30

GOOD STRESSFUL

Once Novartis made the decision to launch phase II, the mood at the company surrounding STI-571 changed completely. The stalling had been replaced with a blank check and a cleared schedule. "When Novartis got behind this project, [it] was a high-speed train and you just got out of the way," said Druker, who credits Vasella with the company's turnaround. No expense was spared to produce sufficient amounts of the drug as quickly as possible. Production of the drug was moved from Basel, where only small batches of STI-571 could be made at a time, to its manufacturing plant in Ringaskiddy, a village in County Cork, in southern Ireland. Ringaskiddy had only been used to produce already approved drugs, never anything in the pipeline. To manufacture hundreds of kilograms of a completely new compound as quickly as possible, the plant was pushed to its maximum capacity. The team there was working twenty-four hours a day, seven days a week.

As Druker had predicted, the presentation of the phase I results at the 1999 ASH meeting had created an enormous expectation. Suddenly, "everything has to be done by tomorrow," recalled Renaud Capdeville, a pediatric hematologist from France who'd joined Novartis in 1997 and taken over command of the STI-571 clinical trials in 2000.

The decision to expedite production of STI-571 was driven by the need to open the trials to as many patients as possible. But once the company had committed itself to developing the drug, another pressing need quickly followed: getting it reviewed by the FDA as soon as possible. Although it does not regulate how doctors use the medications it approves, the FDA does regulate how products are labeled and sold. An FDA approval grants a pharmaceutical company permission to advertise a drug for the indication specified in that approval. Pharmaceutical companies cannot legally market their products until they are approved by the agency. All of the data from the clinical trials of STI-571 would have to be sent to the FDA's Center for Drug Evaluation and Research for review. In the late 1990s and early 2000s, the FDA had been accused of taking too long to review beneficial new medications, and Novartis knew that the road to approval could be a long one. Clinical trials were all about investment, and these studies were costing the company hundreds of millions of dollars. Not a single dollar could be earned from STI-571 until that approval was granted.

Despite the company's commitment to developing STI-571, the rarity of CML was still an issue. STI-571 would be a patented drug, and eventually that patent would expire. When the patent expired, so would the profits. "When you lose patent, basically 90 percent of the volume is gone within a few days," said Vasella. "It just evaporates." Once the patent expired, generic manufacturers could use the drug formula to make their own capsules. With STI-571, the patent pressure was intensified because the number of prescriptions during its proprietary years would be fewer than usual. The rarity of CML meant the volume of sales would be far lower than for the average medication coming through the clinical trial system. STI-571 wasn't even approved yet, but there was no time to wait on strategizing how to wring the most money out of it. The first part of the plan was to get the drug approved as soon as possible, so that Novartis could maximize sales while it had exclusive rights to the drug formula.

STI-571 wasn't the first such drug to come through the trials system, and the FDA had mechanisms in place for drugs that addressed an unmet need, as STI-571 did. Before the phase II trial had launched, the FDA had granted STI-571 fast-track designation and accelerated approval status. Normally, after the small phase I and larger phase II studies, a new drug must go through a phase III trial in which it is directly compared to the current best treatment option before the FDA would consider approval. For STI-571, that would have meant a large study in which patients were randomly assigned to treatment with either the experimental drug or interferon plus ara-C, and their responses would be compared after several years. The data were too promising to wait that long. The current best treatment was not all that great, and holding up approval of STI-571 until completion of a phase III study would have been unethical.

The fast-track designation gave the company enhanced access to the FDA—more meetings, more correspondence, more assistance—to expedite the clinical trials process. Because STI-571 gave CML patients a new and necessary treatment option, the FDA was willing to take the time to give Novartis careful guidance about exactly what would be needed to prepare for approval review.

Accelerated approval meant the company could submit its phase II data for FDA review, with the agreement that a phase III study would be completed if the drug was approved. Because the phase I and II trials are short, single-arm studies—that is, all patients received the same treatment, rather than being randomized to one of two arms, as they would be in phase III—the results don't hold as much clout. But the advantage of submitting phase II data as opposed to phase III data was quite significant. It shaved years off of the development time. With accelerated approval status, the FDA would review STI-571 based on the surrogate end points of hematologic and cytogenetic response. Changes in blood counts and the amount of cells with the Philadelphia chromosome didn't prove that the drug worked, but they did point strongly in that direction. Accelerated approval status meant those measures, alongside the data on side effects, were sufficient for an FDA review. Conditional approval granted all the same marketing freedoms as a standard approval.

Even with this special status, Novartis still needed a year's worth of phase II data before submitting the drug for approval. Novartis—in particular, Jörg Reinhardt, the global head of development—was pushing to file the data with the FDA by December 2000, just two and a half years after Bud Romine took his first 25-milligram pill. That timetable was at least a year less than average for drug development. The pills would need to be manufactured quickly.

STI-571 is created through an intensive, multistep process. Following the intricate cascade of chemical reactions that Jürg Zimmermann had plotted out when he had created the compound years earlier, the raw materials are mixed in solvent, heated, evaporated, condensed, filtered, and put through several other exacting chemical reactions out of which the powder eventually precipitates. Then the powder has to be milled into pills, which also takes time. As with any serious medication, STI-571 had to be created under the strictest quality control, with extra precautions—protective suits and a plantwide containment system—to prevent exposure to hazardous chemicals used to make the drug.

With instructions for making the pills translated from German, the manufacturing team at the 60,000-square-foot plant in Ringaskiddy was ready to begin production. By November, the first seven of the twelve steps required to produce STI-571 were complete. For the remaining five steps, which would normally take a year, Novartis gave the plant seven months. The company wanted to have a metric ton of material by the summer of 2000. To produce that volume, the Ringaskiddy plant had to be operating twenty-four hours a day, seven days a week, its entire staff put on the job.

The phase II trials for blast crisis patients opened in August 1999, and continued accruing patients until March 2000. The phase II study for patients with accelerated disease, those with slightly longer to live, opened the same month and was full by April 2000. The phase II trial for chronic-stage CML patients began enrolling in December 1999.

By the end of 1999, trial sites had opened at Hammersmith Hospital, in London, with John Goldman as the lead investigator; at Druker's former workplace, Dana-Farber Cancer Institute; and at cancer centers in Detroit, Miami, and New York. By early 2000, nineteen sites were up and running. In early January, Michael O'Dwyer, an Irish doctor who'd come to OHSU to study hematology, joined Druker's group to help with the ongoing clinical trials there and Druker's continued research of CML and tyrosine kinase inhibition.

On January 1, 2000, Suzan McNamara, the woman behind the Internet petition, flew to Portland, Oregon. As soon as she arrived, she went to OHSU for preliminary testing. She could barely walk up the stairs. She was 5' 8" tall and weighed 110 pounds. After reviewing her test results, the staff told her that they thought she was just at the borderline of accelerated stage, no longer in the early, chronic phase of the disease. The thought panicked her because it meant that, even though she could enroll in the trial, the disease might already be too advanced for the drug to have any long-lasting benefit. When she returned to the hotel, her phone was blinking with a message. It was Druker. "Suzan, I don't want you to worry," he told her. "Your tests came out fine. We're getting you on the trial."

Within a week, after suffering through a brief but sharp bout of leg pain, McNamara was already feeling better. She stayed in Portland for five weeks with her boyfriend. "We had the time of our lives," she says. "I felt reborn." Among her best memories of those weeks is all the eating she did. For the first time in two years, she could eat whatever she wanted. By the time she left Portland, she had gained twenty pounds.

But even with the company's full commitment, the drug supply still fell short. For the phase II study of chronic-phase patients, which had opened in December 1999, the shortage was all too real. Each trial site was allowed to enroll ten eligible patients per month. But Druker already had a waiting list of 200 patients when the study first opened. How could he, or any of the other investigators, who also had waiting lists, choose which ten to enroll in a given month? "I was having to prioritize patients and make decisions about who should get this drug and who shouldn't," said Druker. People were showing up at OHSU to demand access. He never turned anyone away from the study; he just added names to the list and tried to not have anyone wait longer than three or four months. For some, that wait would be too long, and they would die of the disease. As usual, to relieve himself from the physical and mental stress, Druker ran. In August 1999, it was the Hood to Coast relay, stretching from Mount Hood to the Pacific coast beach. In October 1999, it was the Portland Marathon. The rest of the time, it was the steep hill that led to his clinic.

He also spent time staring at his office wall. When the phase I patients started responding to STI-571, Druker asked his group to send him photos of themselves doing the things they loved to do. "Let's make this believable," he told his patients. "Give me pictures I can show." He tacked them up on his office wall as evidence to any visitor of what was happening. A patient who'd entered the study crippled by fatigue was gardening. Judy Orem sent a photo that showed her planting a tree. Another was surrounded by her grandchildren. Stuck once again with an excruciating wait, the wall gave Druker the will to persevere.

Finally, as promised, by the summer of 2000, enough of the drug had been produced to support full enrollment of the phase II study for chronic-stage patients. At last, the floodgates opened. And patients rushed through. The phase II study for chronic phase patients—those who typically had four to six years to live—enrolled 532. The phase II study for accelerated phase patients—with about one to two years to live—enrolled 235. The phase II for blast crisis patients—who had just months to live—enrolled 260. "Could it have happened faster? Sure," Sawyers acknowledged. But Novartis "did the right thing in the end."

SINCE DRUKER'S ARRIVAL in 1993, OHSU had gone from having a small cancer center where only people from the area sought care to a hub of leukemia care, with patients flying in from around the country and the world. With the three phase II studies now fully open, it was time to add another doctor to the overtaxed clinical team, which still consisted of Carolyn Blasdel, two other research nurses, and Michael O'Dwyer. Druker needed another clinician to handle the patient load, and Michael Mauro was the ideal candidate. A medical doctor just finishing up a fellowship at New York Presbyterian, Mauro was trying to figure out his next steps when he saw a job posting for a clinical hematologist. He knew of Druker's work and had been in the audience at the 1999 plenary session at ASH. Mauro assumed that there would be scores of people like him vying for the chance to work with Druker on these trials, and was shocked to receive a call from Druker, who wanted to interview him for the position. "He just wanted to talk," recalled Mauro. They discussed leukemia research, Mauro's experience, and running, an interest they shared. He got the job.

By July 2000, Mauro and his wife, a designer, had moved to Portland. With barely a moment to take in his new surroundings, Mauro was immersed in the clinic. From the start, he was seeing trial patients, consulting with new patients about trials they could enroll in near their homes, monitoring all the test results for the clinical trials, and accepting invitations to lecture across the country and internationally. "And this was all in a matter of the first year or two," said Mauro, a lithe man with wavy brown hair and green eyes. Less than two months after starting, he was submitting research to ASH. Like Druker and the rest of the team, he worked nights and weekends, doing everything he could to care for patients and spread the word about the drug. The relentless schedule was, as he puts it, "good stressful," Mauro said. "I wouldn't trade it for the world."

Mauro already knew the unprecedented physical responses to STI-571 but wasn't expecting the emotional response that patients had for it. The patients treated their medicine like gold. "They cherished it, they defended it with their life." Mauro remembered one couple saying they'd taken the jewelry out of their safe to make room for the bottles of pills.

As Druker, Sawyers, and Talpaz had, Mauro began amassing his own inspiring stories. One patient he met early on had developed delirium from interferon and was being treated for a mental illness as well as cancer. She started taking STI-571 and underwent a complete reawakening. She returned to living independently and to being with her children and grandchildren.

The phase III trial did not lag far behind. Dubbed IRIS, short for "International Randomized Study of Interferon and STI-571," this study was the first one to allow newly diagnosed patients who had not yet started any other treatment for CML access to STI-571. Even though the company could submit the drug for FDA review based on phase II, the randomized study comparing STI-571 with interferon had to get started. Novartis wanted the drug available to newly diagnosed patients as soon as possible if the evidence supported that use. The clinicians were equally eager because they knew that patients should have access to STI-571 the moment they'd been diagnosed, not after they'd suffered through interferon. By the fall of 2000, Novartis had the still-ongoing phase I trial, three phase II studies, and a large phase III trial open to enrollment of just over 1,000 patients.

But there was still more to go. After the phase II studies had reached their target accrual numbers, plenty of CML patients around the world were still taking interferon. Every CML patient whose disease had progressed on interferon wanted the drug. STI-571 was no longer considered an altruistic option to advance medical knowledge. It was a medication that was almost certain to help. But these patients would not be able to get into the phase III study because it was reserved for patients who had not taken any other medications for CML. Yet they couldn't get in the phase II because all the study sites were full. Sawyers remembers returning to the office after a long day at the clinic to a stack of messages from people around the world wanting to get the drug. "You felt that it was your obligation to get them some [STI-571] because you knew it would help them," he said.

Toward the end of 2000, the company launched yet another trial, this one called "expanded access." This study allowed individuals with CML who'd stopped responding to interferon or could not take it at all and were ineligible for a bone marrow transplant to get STI-571 through one of the phase II investigators without enrolling in one of the previously established trials. The process was laborious; the paperwork required for each patient amounted to a mini–clinical trial for each enrollment. But now, almost any CML patient who wanted the drug could get it. The only restrictions were distance—patients still had to be willing to travel to one of the trial sites periodically—and the long amount of time it took for the staff to enroll each patient; they seemed to always be behind.

By the end of 2000, trial centers existed in twenty-eight centers across Europe and the United States—fourteen countries overall—and more than 3,000 CML patients were being treated with STI-571 worldwide. People from countries without an official trial site traveled long distances for an initial appointment with a study investigator, continuing treatment under the care of their local oncologist. The waiting lists were shrinking, and the responses were lasting. With more than 70,000 CML sufferers worldwide, there were still plenty of patients waiting to benefit from the drug, but the drug was still experimental. Some conservative-minded patients and doctors preferred to wait for FDA approval. Others still hadn't heard about it. But by and large, STI-571 was making its way to those who wanted it.

31

PUTTING IT IN WRITING

In late 1999, Judy Orem started the *STI Gazette*, a monthly newsletter that served as yet another glue for the OHSU phase I trial group. Orem was continuing to respond to the drug. She and her husband had decided to stay in Portland, where they'd both grown up, after her first three-month clinical trial commitment finished. But with others returning to their homes, the clubhouse atmosphere of the OHSU phase I group started to wane. "It was a sudden loss," said Orem, who'd grown accustomed to long conversations with patients who'd started the trial around the same time she did.

The *Gazette* was a way to keep the information flowing and the family spirit alive. "I sent it out every month so that people, when they left the study, still would feel a part of it," she said. Orem had taken on a kind of mother-hen role, organizing events and making sure people knew the latest news about the drug. Each issue had a personal message from Druker or another doctor with insights about how STI-571 had come to be, the expansion of the trials program, responses to questions from patients, and, later, the investigation of the drug in the treatment of other cancers. The newsletter, paid for and printed by the Leukemia and Lymphoma Society, was also a place to collect patients' stories and updates. By now there was a growing sense among the phase I enrollees that they were in the midst of something special. That was true for Druker's patients, but also for people around the world who, through the newsletter, could express their sense that they were part of something bigger than themselves. Sending in a story to the newsletter wasn't just a way to communicate how meaningful the drug was to each individual. It was a way to document a miracle.

A patient named Dori wrote of her week spent hiking in eastern Oregon. Another patient announced a long-awaited return to her pottery work. One issue included a recipe, another joked about a tattoo revealed during a bone marrow biopsy, and yet another wondered about the owner of an Arizona license plate that read "STI-571." There was information about organizations that provided cheap or free flights to cancer patients needing to reach faraway treatment centers, other practical advice and insights, and an ongoing chronicle of the progress of the drug through the clinical trials system. Each otherwise mundane tidbit shared was tinged with the awareness that it would not have happened if not for the drug.

Then there were the personal histories. The newsletter printed first names only, but everyone knew who was who. Sandy, from London, had been diagnosed when she was already in the accelerated stage of the disease. She was about to get a bone marrow transplant when she was accepted on a trial. Three months later, the disease had reverted to the chronic stage.

Gerry had been diagnosed after inexplicable bruises suddenly appeared on his thighs. His white blood cells were 100 percent positive for the Philadelphia chromosome when he was diagnosed. After three years on interferon, he slept twenty-two hours a day. He entered the STI-571 trial as part of the 85-milligram cohort and returned later to treatment with 300, then 400, then 600 milligrams. The drug was working. "It's going great guns," he wrote. His white blood cells and platelets were under control, and the percentage of Philadelphia chromosome–positive cells was steadily diminishing.

Rose was taking chemotherapy and interferon for CML when she heard about the trial on television. "Leukemia didn't make me sick, just the medicines," she wrote in the *Gazette*. By July 1999, after four months on STI-571, her blood and bone marrow were clear of the mutant gene. High medical bills and her inability to work had left her destitute. She and her husband lost their house and car and filed for bankruptcy. But, she wrote, "Now I can look forward to playing with my grandchildren."

Lucille entered blast crisis in July 1999, four years after being diagnosed with CML during a routine checkup. She assumed she'd be dead within six months when she started STI-571. Three months later, she was traveling to her daughter for Thanksgiving to see her and her own grandchildren and great-granddaughter. "I'm feeling well," she wrote. "I have so much to be thankful for."

Linda's husband was on the phase I trial at OHSU, and she wrote about how the drug had changed her life. "Now I can get really pissed off at my kids . . . without being concerned that I should be a bit more lenient because their father has cancer. . . . I can look at photos from a trip we took to London before the diagnosis—and not start crying. . . . I can stop calculating survival statistics."

Mark's wife was diagnosed in 1995. Interferon worked wonders for a while, though it left her tired and with flu-like symptoms lasting four years, including a persistent fever of 102 degrees. He'd cut down on work to help care for their children, their home, and his wife, including giving her the painful injections of medicine each day. He had read about STI-571 online, and they'd met with Druker when interferon was still working, keeping her white blood cell counts in check, though she felt miserable. Eventually, she developed heart damage and had to stop the interferon. She enrolled in an STI-571 trial, and within a few months her blood work was completely normal.

Eduardo was the first Brazilian to take STI-571. Five months after starting treatment, he was playing tennis and swimming several times a week and had been featured in a Brazilian magazine as "the local STI case," helping to spread the word to other CML patients in his country.

Then there was LaDonna, who wrote her story for the June 2000 issue of the *STI Gazette*: "In February my funeral was arranged. I said goodbye to my children and family. I was very tired of being sick and looked forward to peace in death. I could not tolerate interferon[,] and hydrea did little but give me stomach distress. I couldn't eat or hold anything down." On the way to see LaDonna at the hospital, her husband picked up a local paper. His eyes fell upon an item about Druker and the STI-571 study. On February 16, 2000, LaDonna entered the phase II study for accelerated stage CML patients.

"I came in a wheelchair and could barely walk," LaDonna wrote. "I did not want to do the study. My family and doctor insisted. My Philadelphia chromosome was 97%. My white count was 220,000. My spleen was very painful and large (29 cm). I was on pain patches and couldn't sit up. Four months later, my white count is normal, no blasts, and the Philadelphia chromosome is 1%! I'm able to function and have fun with my family." Her spleen shrank to 6 centimeters, and she was no longer in pain.

Like so many of the profiles, LaDonna's "Message of Hope" ended with gratitude: "Thank goodness for Dr. Druker and STI-571. Every day is a gift from God."

Expressions of thanks to Druker were woven throughout every issue of the *STI Gazette*. "Kiss Druker. Hug Druker. We love you, Dr. Druker," two patients wrote. They swapped ideas for T-shirts bearing his name to wear when they were interviewed, "just to keep this modest humanitarian in everyone's mind as the man who made all this possible," someone wrote. In June 2000, one patient reported that she had planted a tree in his name in the Millennium Forest outside Jerusalem. Again and again, patients thanked him, Carolyn Blasdel, and the rest of the staff. Online, on the message boards, the sentiments were the same. As far as they were concerned, Druker had saved their lives.

With all of these miracle stories, the public began to take an interest. The media attention surrounding the ASH conference of 1999 had brought the drug into the public consciousness. In Oregon, local papers continued covering Druker's work, though national coverage died down as the phase II trial got under way. Then, at the end of 2000, almost a year after the phase II studies had been fully launched, *People* magazine decided to do a profile of the doctor behind the cancer wonder drug. The story was assigned to Alexandra Hardy, the same writer who'd interviewed Druker five years earlier for the Associated Press.

Druker remembered her. He remembered that she had asked him when the clinical trials would start for this compound that had proved so interesting in the laboratory. He also remembered her boots and hair. After the interview for *People*, Hardy and Druker developed a friendship. It turned out that they belonged to the same gym. Hardy's story was published February 19, 2001. A few weeks after that, said Druker, "I got the nerve to ask her out."

Hardy had found Druker to be a tough profile subject. Everyone she interviewed kept praising him, and as a journalist that made her nervous. One of his patients told Hardy that Druker was "right up there, right under God." The stories and accolades were too fawning for a journalistic profile. She searched for one person who could offer some balance and make this doctor seem a bit more real. She never found such a source during her interviews. Afterward, however, she did. "I became that person," she said. As they fell for each other, this doctor who was so idolized by his patients became human to her. Finally he had someone who could know him outside of the hospital, who could see him not just as a doctor but as a regular, imperfect person—a man who often lacks diplomacy when expressing his opinions, and who gets cranky when he's hungry—and as she came to see that imperfect person, she found she liked him all the more.

32

A TRUCKLOAD OF DATA

The results seen in phase I continued strongly in phase II. In the study of accelerated stage CML (more than 15 percent blast cells), 149 out of 181 patients had a positive hematologic response. In 96 of those patients, red and white blood cell counts returned to normal. Cells with the Philadelphia chromosome began disappearing. One year after starting treatment, 43 of 181 patients had a major cytogenetic response, with 30 patients showing no sign of the mutant chromosome.

The story for the phase II trial of blast crisis patients (30 percent or more blast cells) was the same. Here, the responses were somewhat less dramatic; these were, after all, the sickest patients, and the investigators knew from the start that even those who responded vigorously to the drug at first could relapse just as fast. Of the 229 patients, 119 had some kind of positive hematologic response. Sixty-four patients returned to chronic-stage CML, and in thirty-five, blood counts normalized completely. The responses lasted an average of ten months, with some lasting more than a year. These were patients who, like LaDonna, had either decided what kind of burial they wanted or were too sick to think about it. About fifteen of the blast crisis patients had no mutant genes when the study concluded.

The phase II trials of blast crisis and accelerated stage patients tested two dose levels, 400 milligrams per day and 600 milligrams per day. Both worked, but when the investigators analyzed the data, they noticed that patients who took the higher dose tended to respond better and live longer. Higher doses, they could see, might be the best approach for patients with more advanced disease.

In the phase II trial of chronic CML patients, 532 patients had taken 400 milligrams of STI-571 for an average of 254 days when the data were analyzed. About 250 patients had a dramatic drop in the number of cells containing the mutant gene. In all but seventeen patients enrolled in the trial, the disease was halted.

In the expanded-access study, a widespread collection of individual clinical trials for patients who were still in need of the drug after the phase II trials had completed enrollment, the responses followed the same patterns. The results of these three studies were without precedent in the history of cancer treatment. No drug had done better.

The side effects were the same as in the phase I study. There was some nausea and some diarrhea, but those subsided after patients acclimated to the medication. The stomach cramps let up, and the leg pain ran its course. Puffy eyes abounded. There were some rashes here and there. One patient died from liver toxicity, most likely because the presence of large amounts of acetaminophen in that patient's body had exacerbated an otherwise minor problem, an unforeseen complication. Between 10 and 20 percent of patients did experience a severe drop in blood counts, a serious condition that may have been a sign that the drug was working, since the mutant Philadelphia chromosome had been triggering the growth of too many white blood cells.

By the end of 2000, Novartis had enough data to submit the drug for FDA review.

THE REPORT SUBMITTED to the FDA included all of the toxicology studies, the phase I trial, and the phase II trials. In addition, the FDA was given all the information about the composition and manufacture of the drug: what chemicals were in it, the step-by-step process of how it was prepared, all other particulars about the manufacturing process, its inactive ingredients, and its form.

In early February 2001, a truck carrying boxes of binders filled with pages upon pages of data and information left Novartis. On February

27, 2001, the drug was officially filed for review by the FDA. Forty years had passed since Nowell and Hungerford spotted the abnormally small chromosome, and thirty years since Janet Rowley discovered the translocation at the heart of the Philadelphia chromosome. The binders sent to the FDA were the culmination of decades of research—the discovery of the Abelson virus at an NIH laboratory; the piecing together of the Bcr/Abl fusion protein and its connection to CML by Owen Witte, Naomi Rosenberg, and many others; the gradual shedding of light on kinases and phosphorylated tyrosine; the cellular origin of oncogenes; and proof that the mutant gene encoded a mutant kinase that was the sole cause of CML. Eighteen years had passed since Ciba-Geigy gave Matter the resources for his under-the-radar kinase inhibitor program. Years of chemistry work to create a molecule that could block that kinase had eventually generated an experimental molecule that blocked the kinase in question, and years of struggle to turn that molecule into a drug had followed. For long stretches, no one involved in the work felt certain this moment would ever come, and yet they'd all held out hope that it would.

Finally, the principle had been proved. The kinase had been killed, and so had the cancer. Now the future of this drug was in the hands of the FDA. So much rested on the approval of this drug—the lives of CML patients, and possibly, if the principle applied to cancer generally, the lives of so many other sufferers. Proving the principle of kinase inhibition had come to be about much more than treating a rare leukemia. The drug was changing the way people thought about cancer. This deadly illness, so impervious to decades of medical efforts, was finally being wrestled into a new era by virtue of the growing clarity that cancer is, at its root, a genetic disease. With that knowledge, the possibilities for bettering treatment were not just exciting—they were real and concrete, backed by evidence, grounded in logic and science.

When a new drug application is filed with the FDA, the application doesn't arrive as a surprise. The agency is well aware of the progress of various experimental drugs and the status of submissions for its review. In this case, the FDA was especially prepared. Agency officials knew about the study results so far and the media coverage of the drug. The fast track and accelerated approval designations meant the agency had been involved more closely than it usually was with new drugs.

Novartis had also secured another FDA designation for STI-571 that garnered it special attention. Just days before Novartis shipped off the binders full of clinical trial data, the FDA granted STI-571 orphan drug status. Orphan drugs treat rare diseases, those occurring in fewer than 200,000 people per year in the United States. Orphan drug status can also be granted to drugs that treat more widespread diseases but that, for one reason or another, are unlikely to recoup their development costs. For the bulk of pharmaceutical industry history, rare diseases were largely ignored because companies knew that the money required to create a drug was unlikely to be recovered and that profits were even more unlikely. In addition, conducting clinical trials for rare diseases was thought to be exceedingly difficult because there wouldn't be enough patients for a trial to yield statistically meaningful data.

The Orphan Drug Act, or ODA, was passed in 1983 to change this dire circumstance. The law provides incentives for the pharmaceutical industry to turn its attention to orphan ailments: federal funding for clinical trials, a 50 percent tax credit on trial costs, and, most important, seven years of market exclusivity. That last provision guarantees that no competing drugs will be sold for the same illness for seven years, even if the patent expires. Patents are obtained much earlier in the development process, often before an experimental compound has been made into a drug. This protection generally lasts for thirteen years, but by the time clinical trials are completed and the FDA approves the drug, those years may be more than half over. Market exclusivity, which is not granted to drugs for more common diseases, means that a company with a successful new drug for a rare disease will own that market for a full seven years.

The ODA has dramatically influenced the pharmaceutical industry's willingness to add rare diseases to its scope of interest. Before the act, there were ten drugs approved for the treatment of rare diseases in the United States. By 2010, 367 orphan drugs were on the market, with more than 2,000 in the pipeline. The federal financial incentives, coupled with the high prices typically set for these drugs (some cost

$500,000 or more per year) turned rare diseases into a goldmine for the pharmaceutical industry. The orphan drug market was worth more than $58 billion by 2006, with a predicted annual growth rate of 6 percent, bringing that amount to more than $112 billion by 2014. Even if an orphan drug later turns out to be a blockbuster (as happened with Botox, originally approved for the treatment of a rare muscle disease and later approved as a wrinkle smoother), the drug maker is assured that the ODA benefits will not be rescinded.

Because CML struck only about 5,000 people per year in the United States and treatment options were so inadequate, STI-571 was a perfect candidate for orphan drug status. The designation didn't ensure profits, but it did take a significant edge off the investment. The tax credit and market exclusivity would certainly go a long way toward paying back the company's investment. And its standing as an orphan drug made reviewing STI-571 even more of a priority for the FDA.

In March 2001, with the data already under review, Novartis received the news that STI-571 had been granted priority review. This status completed the process of special consideration that had begun with fast track designation, which gave Novartis privileged access to the FDA, followed by accelerated approval status, which enabled the drug to be reviewed based on the surrogate markers of benefit measured in the phase II studies. Priority review simply meant that the FDA would read through all the data more quickly than it might for a drug that was not offering such a significant improvement over currently available treatment options. The goal was to have an answer by six months after the submission date for the new drug application. The Novartis executives, the investigators, and the patients were anticipating that the drug would be approved by around September 2001.

With the submission, Novartis was required to provide a name for the drug. The generic name the team had selected was imatinib mesylate. The brand name chosen by the company was Glivec, pronounced "GLEE-vek."

Under fast track and accelerated approval, Novartis had worked with the FDA to craft the phase II studies in a way that satisfied the agency's criteria for reviewing new drugs: what kind of response could be considered a true surrogate for clinical benefit, what defined an interferon failure, how much follow-up time was needed for the patients in the trial.

The review period was fairly uneventful for a while. The agency had some questions and so requested more information. Some wording in the proposed package insert needed changing, and the company had to clarify its definitions for certain aspects of the disease or response to the drug. Then more wording changes for the package insert. Mostly the issues were minor and easily resolved.

Of all the information included in those binders shipped to the FDA, the name proved to be the greatest sticking point. Names draw heavy consideration in FDA new drug reviews. A proprietary name is not allowed to sound like the disease it treats, and it can't be too similar to other drugs on the market, even if those medicines address vastly different problems. The choice of brand name follows no obvious metric. Generic names tend to be garbles of mysterious syllables, occasionally referring to the type of molecule at hand (the "inib" in imatinib, for example, is short for "inhibitor"). By comparison, brand names are easier to remember—Glivec, instead of imatinib mesylate, certainly fit that bill. But the goal of creating a memorable brand name doesn't explain where the names actually come from. Often, pharmaceutical marketing teams select from lists of potential names, plucking one at random that seems to fit. Novartis's choice had nothing to do with happiness. It was just a name.

In mid-April, after an extensive review of the proposed proprietary name, the agency told Novartis that its choice, Glivec, sounded too similar to a drug named Glyset. Novartis fought the decision. The medication was for a serious, rare disease, and therefore would be distributed in a highly controlled manner to a very small population. It wouldn't be crossing paths with Glyset, so mixing up the drugs wasn't an issue. The Glivec pills in no way resembled Glyset, a diabetes drug. Plus the diabetes drug name was pronounced with a long *i*, to rhyme with *cry*, not with a long *e*, as in *glee*.

The FDA was not swayed. That more than one hundred cases of mix-ups had been reported among the drugs Celebrex, Celexa, and Cerebyx, each for different uses, was proof that the sound-alike concern was well founded. The proprietary name study and ensuing debate runs fifteen pages long in the FDA's review document. After the prolonged debate, the solution was simple. The pharmacist conducting the name review suggested that the company change the name to Gleevec, to match the pronunciation of Glivec. Novartis relented.

Many trial patients, especially those who'd been there from the start, took an immediate dislike to the name. They aired their complaints in the *STI Gazette*. To them, the name "Gleevec" sounded totally strange. "Seems a bit odd for such a killer drug," one patient wrote in, "but who's complaining." Some wanted it to be named Drukercillin. To all of them, it remained STI-571. Referring to it by the pipeline name, or even just STI, like a nickname, became a point of pride and sentiment. Those who were truly in the know and whose lives had been saved by this little pill would always refer to it as STI-571.

On April 30, 2001, the FDA sent a fax to Novartis requesting that the proposed package insert—the folded paper of fine print that would be included with the bottle of pills, should the drug be approved—include all of the side effects that had been observed in patients during the studies, not only those thought to be caused by the drug. Novartis knew that, per the accelerated approval designations, they were guaranteed an answer from the FDA regarding approval by September, six months after the data had been submitted. Communications from the FDA were only a sign that the review was continuing, not that approval was imminent. Most drug reviews, those that were not accelerated, took twelve to fifteen months on average. The goal of a six-month review seemed fitting for a home-run drug like STI-571. But was it possible for the agency to move that fast?

On May 10, 2001, Novartis received a letter from the FDA. "We have completed review of this application, as amended, according to the regulation for accelerated approval," it read, "and have concluded that adequate information has been presented to approve Gleevec (imatinib mesylate) 50 mg and 100 mg capsules for use as recommended."

It was the letter that patients, investigators, and various teams at Novartis had been waiting for, some for many years. From a genetic mutation to its haywire fusion protein, from the fusion protein to the leukemia, a path had been traced for the rational design of this drug.

From CGP-57148B to STI-571 and finally to imatinib mesylate and Gleevec, the world's first drug targeted against a specific mutant protein that stopped cancer in its tracks was ready for wide release. The drug was approved.

The FDA approval regulated only how the drug could be marketed. But marketing the drug was exactly what the company had been waiting to do. Now it could sell the drug rather than only pay for its use in clinical trials. Insurers would add it to their formularies, the lists of medications covered by their policies, the final step needed to earn money from prescriptions written to CML patients. Novartis still had to complete the phase III clinical trial; per the accelerated approval program, the FDA's nod was still conditional as the data from the large, randomized study matured and the actual survival benefit emerged. But there was no reason to doubt that anything would change with this drug. Doctors were free to prescribe it. Patients across the country could get it. The drug had been set free from clinical trials.

With the approval, the fears about toxicity shriveled. The FDA confirmed that although liver and kidney problems had been observed during the trials, these were temporary, minor, and reversible. The reviewers concluded that the toxicology studies supported approval of the drug. The package insert did include a warning that women should not become pregnant or breast-feed while on the medication, in light of evidence that the chemical had seeped into the milk of lactating rats. But all of the concerns about what harm the drug might cause—the worries that had spurred tests in rats, rabbits, mice, and monkeys—had been allayed. Even the blood clots in dogs, the problem that had screeched the drug's development to a halt, fortuitously enabling the oral formulation to take center stage, had turned out to be a false alarm. Some time after those canine tests it had become clear that the catheters, not the drug, had been the problem. The last sighs of relief over potential problems had been breathed long ago. The drug was powerfully effective, and it was safe.

33

THE FATHERS OF VICTORY

The approval was announced at a news conference held in Washington, DC, by the National Cancer Institute and the FDA. On May 10, 2001, several agency officials stood in front of members of the press and invited guests to praise those responsible for the breakthrough, and to highlight the rapidity with which this life saving drug had been brought to the people who needed it. The speakers included Tommy Thompson, secretary of the Department of Health and Human Services; Richard Klausner, director of the National Cancer Institute; Richard Pazdur, who directed the oncology drug products division at the FDA; Suzanne Dreger, a patient from Falls Church, Virginia, who'd been part of the clinical trials at OHSU; and Daniel Vasella.

Although Gleevec had already made the pages and websites of every major news outlet, the press conference was the official coming-out party for molecularly targeted medicine. "This drug has been engineered in the laboratory to target a single, cancer-causing protein, and like a light switch, turn off its signal to produce leukemia cells," said Thompson, reading from his prepared notes. "We believe such targeting is the wave of the future." Klausner, the most expert oncologist among the speakers, offered a similar view of the significance of the approval. "This new drug, we believe, is the picture of the future of cancer treatment, and a vindication of the scientific approach to disease," he said.

Everyone acknowledged that the long-term benefit of Gleevec was still unknown. Not enough time had passed to confirm whether the drug actually prolonged the lives of people with CML without eventually causing intolerable side effects. People newly diagnosed with the disease had four to six years to live, on average, and some much more than that. The phase I patients had been on the drug for less than three years, and the phase III study, in which newly diagnosed patients were receiving Gleevec as their first treatment, was still ongoing. So the answer to that ultimate question—did Gleevec prolong life?—was still years away. But, insisted Klausner, molecularly targeted therapies were the key to "a long but hopefully more successful war on cancer."

Klausner differentiated between other drugs that target specific, known substances in cancer cells. He was well aware that the breast cancer drug tamoxifen was really the first targeted drug because it works by specifically blocking the production of estrogen. The difference, said Klausner, responding to a question from a journalist, is that the estrogen receptor targeted by tamoxifen is not the root cause of cancer. CML is a one-gene disease. A single mutant gene encodes a single mutant enzyme, and this enzyme alone causes the cancer. "This was a target that is not just in a cancer but is responsible for the cancer," he said.

His optimism about molecularly targeted drugs extended far beyond CML. "There are scores of drugs now that fit into this class," he said. As the clinical trials of STI-571 had been churning along, the oncology community had been digging for potential molecular targets in other types of cancer. Researchers probed for genetic mutations hidden inside breast tumors, lung tumors, kidney tumors—any malignancy at all. They measured the amounts of different enzymes and other proteins to see if any were occurring in excess, a potential sign that the substance was part of the ecology enabling the lesion to grow, if not the root cause. In academic laboratories, researchers continued to unpack the signaling pathways triggered by a particular gene that may be responsible for causing cancer. By the time the drug was approved, more than sixty potential targets had been identified in breast cancer. Molecular biology had transformed cancer research, and this drug's power was a testament to the promise of this new direction. "It's very difficult to describe what a different world this is than just five to ten years ago," said Klausner. In fact, he emphasized, this new, targeted drug paradigm applied to all diseases, not just cancer.

The speakers repeatedly highlighted the role that collaboration among academia, industry, and government had played in the speedy study and approval of Gleevec. When asked by a journalist how long the drug had been in development, Vasella's response was "2.7 years." He was technically correct: The phase I trial had begun in June 1998, and the drug had garnered approval in May 2001, less than three years later, an extraordinary accomplishment.

Unmentioned, however, was the fact that the trial had started more than two years after the preclinical studies, and that the tyrosine kinase inhibitor project had actually begun in the mid-1980s. So not everyone shared Vasella's view. "It took eighteen years, which is ridiculous for such a project," said Alex Matter. But Matter's estimate includes the years of toxicology studies and the many years it took to create the compound. That extended timetable was now a chronicle of the broad changes in thinking that had accompanied the development of Gleevec: the unwillingness of industry representatives and the oncology community at large to give credence to the notion of targeting a single kinase with a compound synthesized in a laboratory, the slow acceptance of dedicating resources to treat a rare disease, and finally the breathtaking clip of the studies once the drug maker had fully committed to the mission. Those eighteen years traced a new idea from birth through adolescence, with all the accompanying struggles, heartbreaks, and achievements. In eighteen years, a vision had been wrestled into reality.

If there was any remaining tension in the aftermath of approval, it was between Druker and Novartis. Druker's name had been mentioned only in passing at the approval news conference. Now, Vasella heard rumors that Druker felt shafted because he wasn't being given the recognition he deserved. Vasella was not surprised. He had seen people fight for credit after previous successes, and he believed his experience with Druker to be yet another example. "Each time you have a success, everybody's a father and a mother," Vasella said. The company had not publicly acknowledged the scope of Druker's contribution. But considering the company's financing of the drug—Novartis had covered nearly the entire development cost, save for some grants Druker had received from the Leukemia and Lymphoma Society—and the fact that it had, after all, been created and developed by Novartis, Vasella didn't think such acknowledgment was due. And, because he'd joined the kinase inhibition effort in the middle of the toxicology studies, he was unaware of all the work that had preceded it. "For me, he's an unknown," Vasella said.

Druker readily acknowledges Novartis's ownership of the drug. "They made the drug and did the toxicology testing. They financed the clinical trials," he said. "So from their perspective, it's theirs." He has a harder time with the notion that the drug would have been made even in the absence of his efforts. "They never would have done the clinical trials," he said.

Miffed as Vasella was by the whispers that Novartis was not giving enough credit to Druker, others were equally puzzled by how much credit Vasella was taking for himself and Novartis. Shortly after the approval, Vasella wrote *Magic Cancer Bullet,* a book chronicling the development of Gleevec (coauthored with a named ghostwriter, Robert Slater). Druker was interviewed extensively, though the book makes little mention of his efforts over the years. Nick Lydon declined to participate in interviews for the book because he considered it to be a sham. "If it weren't for Brian's efforts, the compound would likely have disappeared," said Lydon. "If anybody at [the company] should have written a book, it was Alex Matter. He had been involved from the start and deserves a lot of credit for supporting the kinase program through many years." For his part, Matter was not chagrined to see Vasella's take on the story. "Victory has a lot of fathers," says Matter, "and so this is fine." John Goldman, who had to wait for two years after the phase I trial opened to give the drug to his patients in the United Kingdom, holds a similar perspective. "Without Vasella, Novartis would not have made the drug," he said. But the glory reaped for Novartis and Vasella went beyond what was justified. "I don't think it was inappropriate," he said of the credit Vasella and Novartis had claimed in *Magic Cancer Bullet* and in general. "It was just sort of disproportionate."

Druker continued to work with Novartis, tracking the progress of the patients from the phase I, II, and III trials. In 2008, he received funding from Novartis that enabled him to establish one of the first molecular testing facilities in the country. Vasella and Druker have not worked directly together since the approval of Gleevec. They met for the first time in 2009, at the ceremony for the Lasker Award, the highest honor in the field of cancer research, given jointly to Druker, Sawyers, and Lydon that year. Although OHSU has served as a trial site for other Novartis drugs, Druker has not been part of their development. "At times, I'm at peace with my relationship with Novartis. We have a drug that's helping lots of people. That's great," said Druker. "Could they have been better early on and gotten an earlier start, and could there have been a simpler path to clinical trials? Sure. Did they make up the time? Yeah, they probably did. Have they been gracious afterwards? No, they really haven't. Have they looked at what we're doing and said, 'We did this once, why can't we do this again?' No. But again, I'm at peace with that, and that's their business. They do their things their way, and I have to do what I need to do."

For Druker, the actual moment of the approval was one more gust in the whirlwind of 2001. His passport from that year marks the trail of talks he was giving around the world—Italy, Germany, Australia, Japan. Oncologists everywhere wanted to know all they could about the drug. They wanted to learn how to give it to their patients, what kind of monitoring was needed, how to decide on the exact dose, what information patients needed to take care of themselves outside of the clinic. After all, the idea of taking a cancer medication at home was relatively new. Most drugs were given intravenously in hours-long injections at the clinic, where nurses were standing by, ready to address the side effects that were bound to arise. With Gleevec, patients were away from professional supervision, and doctors needed to know what to tell them to watch out for, when to take the medicine, whether to take it on a full or empty stomach. That situation might be normal for the majority of prescription drugs, but cancer had never been treated at home. And oncologists near and far wanted to meet the doctor who they believed was largely responsible for bringing this drug into the world. "I was in pretty high demand and I wasn't very good at saying no," said Druker.

That was also the year that Druker and Hardy fell in love. After all the years of 90-hour work weeks, all the pushing to get the drug made, all the time spent caring for patients on and off clinical trials, Druker now found himself in an entirely new situation. For the first time in his life, he had a family of his own. He and Hardy married the following year. He became a stepfather to her first child, and they would soon have two more children together.

PART 4

Aftermath

Every day that the phase I trial patients swallow their next Gleevec pill tests the notion of treating cancer as a disease driven by genetic abnormalities. For as long as they live, they will be part of an experiment that has transformed cancer care. Their continued health proves the principle behind targeted therapy, bringing hope not only to people stricken with CML but also to the world at large, anxiously awaiting the day when cancer will no longer claim so many victims.

But is the success in treating CML a one-off case, the likes of which are unlikely to be seen again? Or are we indeed standing at the precipice of a new era still in its infancy? Treatments tailored to the genetic drivers of cancer may be the future, but there is no telling how or when that future will take shape.

34

A PRICE TO PAY

With the approval of Gleevec came a price tag. In setting the price, Novartis decided to match the cost of interferon at the time. For the 400-milligrams-per-day dose level, the price was set at $2,000 to $2,400 per month. For 600 milligrams per day, the typical dose for accelerated and blast crisis patients, the price was set at $3,500 per month. For the average CML patient, the annual cost of care—medication plus clinician visits—came to just under $33,000.

The price was not higher than interferon's, and interferon hadn't been criticized for its cost. Still, the Gleevec price drew criticism because so many people couldn't afford it. Medicare's lack of coverage for oral cancer medications didn't help. Until then, cancer treatment was predominantly given by injections, and the formulation of a drug (by needle, by pill, by patch) factored into how much coverage Medicare provided. Cancer drug injections were given at a hospital or clinic under close supervision, and it was that whole system that Medicare covered, with code numbers for each aspect of the treatment. Oral cancer drugs that could be taken at home were something entirely new. Medicare's lag time in covering Gleevec stirred up trouble, generating a short but potent spurt of worry and outrage that CML patients on Medicare would not be able to afford it. That problem was later remedied, although Medicare's infrastructure for oral cancer drugs remains highly convoluted and often requires a period of hefty out-of-pocket contributions. But the issue exacerbated objections to the price. Novartis needed to make sure that everyone who needed Gleevec would get it starting immediately after approval.

The company decided to try a sliding-scale assistance program. Anyone with an annual family income of less than $43,000 could get the drug for free. For those with a family income between $43,000 and $100,000, the total amount paid for Gleevec would stay within 20 percent of the annual income. If the family's annual income was more than $100,000, patients would pay the full price.

The assistance program was innovative and helped a great many people obtain an otherwise unaffordable drug for free in the United States and, as global approvals began to flow, internationally. Yet not all the problems were solved. A patient whose family income landed them just on the 20 percent side of the equation could still end up with medical bills too high to cover. Co-pays were often steep, so that even insured patients felt the pinch.

Despite consternation over the cost, it was much too soon to accuse the company of excessive profits. The question of how much Novartis would earn from the drug was still unanswered. The fact that patients took the pill daily had been a revelation, not only because of how unusual the approach was but also because of the financial windfall such a regimen promised. But the long-term benefit of the drug was still uncertain, and therefore so were the long-term profits. There were patients who'd been on the drug for almost three years, but no one knew what the average duration of life would be. There was no predicting what turn of events the next day might bring, and still no knowledge of how long the responses would last.

For this drug without a history, every day was fresh evidence. The first trials may have been completed, but Gleevec was still an experiment. There were no signs that patients would suddenly take a turn for the worse, but there was no proof that they wouldn't.

And Novartis still had to cover the expense of the drug for many trial patients, who were doing far better than anyone expected based on previous experience with cancer treatment and clinical trials. "We phase I patients are provided the drug for life, right?" Judy Orem recalled asking Vasella around the time of the approval. "Yes," she recalled him replying. "But we never expect phase I patients to live very long." The response was tongue-in-cheek. Vasella was beyond thrilled that the trial patients were doing so well, and he knew that the company had an obligation to cover the cost of the medication for those who had willingly subjected themselves to an untested drug. But no one had factored the possibility of a normal lifespan for the patients into the trial budgets.

A year after the drug's initial approval, it was clear that Novartis had nothing to worry about. In 2002, worldwide sales of Gleevec totaled more than $900 million.

ALTHOUGH THE LEADERS at Novartis may not have recognized Druker's contribution, recognition seemed to come to him from every other direction. Alongside the continuous stream of speaking invitations, Druker has also received a series of prominent awards since the approval of Gleevec, including the Lasker Award (which led to a television interview with Charlie Rose), and the Japan Prize, a science award second only to the Nobel in global prestige, awarded in 2012 to Druker, Lydon, and Janet Rowley. His office is lined with commemorative plaques, pictures, and fancy cups that trace the history of his renown over the past decade. In 2007, Druker was asked to take over as director of OHSU's cancer center.

His investment in the clinical research of Gleevec has not brought him wealth, however; he does not own stock in the drug and, to preempt any trouble with insider information abuse, stopped investing in biotech and pharmaceutical companies in 2000. Druker was paid by Novartis for some of his early Gleevec talks, though he notes that the payments were small and that he always used his own slides, never those prepared by the company (avoiding a common practice that has raised suspicions about biased information being used to influence medical professionals).

But the drug's success has helped pay for his research endeavors. His lab receives annual funding that ranges from $900,000 to $1 million from the Howard Hughes Medical Institute (a portion of which covers Druker's salary). He also has a long-standing grant from the

NIH that amounts to about $200,000 per year for his lab, and research funding of about $250,000 annually as part of a $1 million multilab collaborative grant from the Leukemia and Lymphoma Society that has been ongoing for about fifteen years.

And then there's the gift from Phil Knight, founder of Nike. In 2008, the stock market crash in full effect, Druker received a call from Knight, who said he wanted to donate $100 million to the cancer center at OHSU. "I want to invest in your vision," Knight told him. The gift was one of the largest donations ever given to a US cancer institute. At OHSU, the cancer center was promptly renamed Knight Cancer Institute. Overnight, Druker went from being the head of a modestly successful cancer center to being the head of one of the most well-funded medical institutions in the world.

35

SHOWING A WEAKNESS

Following the approval in 2001, the relentless pace of Gleevec development continued. Novartis still had several clinical trials on the go. The phase III IRIS trial had completed enrollment in January with more than 1,000 patients spread across several countries. The drug was also being given to patients with other types of cancer. CML was the only cancer found to express the mutant Bcr/Abl fusion protein, but the other tyrosine kinases inhibited by Gleevec were present in other malignancies. PDGFR, another tyrosine kinase inhibited by Gleevec, was associated with a type of brain tumor. And researchers had also found that gastrointestinal stromal tumors, a rare type of stomach cancer, are driven by mutated Kit, a kinase against which Druker had screened the first experimental compounds sent to him by Nick Lydon back in 1993. Now the time had come to investigate whether the drug was any good at thwarting cancers operated by these proteins.

George Demetri, from Dana-Farber Cancer Institute, and Charles Blanke, from OHSU, were leading a phase I clinical trial investigating the use of Gleevec for gastrointestinal stromal tumors, or GIST. After the investigators tried the drug on a single patient in March 2000 and the disease slowed down, a phase I study opened in July 2000. By May of the following year, just as Tommy Thompson and Richard Klausner were preparing their remarks for the Gleevec approval news conference, Demetri and Blanke were preparing a presentation for the annual meeting of the American Society of Clinical Oncology, the single largest professional gathering of oncologists worldwide.

The results of that first study were published in the summer of 2002 in the *New England Journal of Medicine.* The outcomes weren't as jaw-droppingly impressive as those seen with CML patients, but the drug worked. None of the 140 patients experienced a complete disappearance of the disease, but more than 50 percent had a partial response, the tumors shrinking by at least half their original size. The drug caused more side effects in these patients, with swelling, nausea, diarrhea, and a rash occurring in many. But as with CML, the side effects were mostly minor, not crippling, though five people had serious hemorrhaging. No other treatment existed for GIST aside from surgery. So Gleevec wasn't just an advance in treatment for GIST; it was the creation of a treatment for GIST.

In time, Demetri and other clinicians would see that GIST patients tended to not respond to Gleevec for as long as CML patients did. Even during the ten months of that initial phase I study, twenty patients became resistant to Gleevec. The drug would be approved for GIST—it was far superior to the previous standard therapy. But why did some become resistant so soon?

While Demetri puzzled over the resistance he observed in GIST patients, Charles Sawyers, one of the three phase I investigators from the CML trials, was still haunted by the sight of the blast crisis patients responding so dramatically to Gleevec and then relapsing just as quickly. Witnessing patients with advanced CML rise from the nearly dead and then crash so spectacularly had driven Sawyers back to the laboratory. Why would patients respond so vigorously and then return to the same condition they'd been in before the treatment? Why did some patients not respond to Gleevec at all, even during the chronic stage of the disease? Sawyers had to know what was going on.

The notion of drug resistance existed at the time, but it had never been studied at a molecular level, and this drug provided a unique opportunity. Because Gleevec worked by binding to Bcr/Abl and shutting it off, Sawyers figured that resistance was ascribable to one of two things: Either the drug was no longer blocking the kinase, or the kinase was no longer the sole trigger for the disease.

Just as Lydon had had to invent a way to screen Zimmermann's experimental compounds for anti-kinase activity, Sawyers now had to invent a way to study resistance at the molecular level. Once again, methods were being invented for a need that had never existed. Although the concepts he was testing were straightforward, conducting the actual experiments was not. Looking at what genes or proteins were activated or inactivated at different moments in the progression of a disease was totally new. There were no pathways in place for investigating the cellular changes that caused cancer to become resistant to a drug. Sawyers was the first explorer to map this uncharted territory.

At this point, there were plenty of patient samples to work with. Using the same kind of gel analyses with which Owen Witte parsed Bcr/Abl nearly twenty years earlier, Sawyers could measure the level of Bcr/Abl in the samples from patients who'd stopped responding to Gleevec. The pattern was "absolutely crystal clear," said Sawyers. The levels were high at the start of treatment, plummeted when the patients were doing better, and then crept up again as patients relapsed. Those high levels were evidence that the drug was failing to inhibit the target.

It was one thing to confirm that the kinase was no longer being blocked in patients who'd become resistant. But why? What was going on that would prevent the drug from doing what it did in so many other patients? Why were patients in more advanced stages of the disease more likely to stop responding to the drug? Why did some chronic-stage patients not respond at all? One possibility was that the drug was being metabolized more quickly in patients who didn't respond compared with those who did. If the chemical was broken down and eliminated from the body too quickly, it might not get a chance to enter the bloodstream and plug the binding site in the malignant cells.

Sawyers had begun tackling resistance while the phase II studies were still underway, and a few months after the drug was approved in 2001, he was ready to publish his first findings. In all of the samples he had studied, Bcr/Abl had been reactivated after its initial shutdown. The kinase had stopped its incessant phosphorylation and then started right back up. He had discovered that in some of the samples, a single amino acid on the enzyme had been changed at an area where a critical bond with the drug was supposed to take place. In others, cells were producing excessive amounts of the already mutant *bcr/abl* gene, leading to excessive amounts of the Bcr/Abl enzyme, more than the kinase inhibitor was able to block. Why those changes occurred was still a mystery, but the brambles obscuring the problem were being cleared.

The timing of his report was an unexpected counterbalance, coming so closely on the heels of one of the most celebrated drug approvals in the history of cancer care, an event that had been lauded as the beginning of a new era for targeted drugs, a hallmark of personalized medicine. Sawyers rightly feared the media's spin on his publication, and begged the *Wall Street Journal* reporter who interviewed him to not turn the phenomenon of resistance into melodrama. "It's amazingly simple and teaches us how to potentially get around it," was the message Sawyers wanted to convey about the mechanisms he'd uncovered. Instead, the headline in the June 22, 2001, issue of the paper read, "Gleevec Shows a Weakness in Fighting Advanced Cancer." Sawyers still sighs about it years later, now from his office at Memorial Sloan-Kettering Cancer Center, where his research focus has shifted to prostate cancer and other solid tumors.

These mechanisms explained acquired resistance, the cessation of a previously encouraging response. But they didn't explain the patients who did not respond to Gleevec at all. Those patients didn't have excessive amounts of *bcr/abl*, and they didn't have the altered amino acid. Plus, these "upfront resistance" patients often had changes in their blood cell counts after taking the drug; they just never had any decrease in the number of cells containing the Philadelphia chromosome. And it was far more common to see upfront resistance among patients with more advanced disease; 48 percent of blast crisis patients did not respond to the drug, not even with a noteworthy change in blood counts. As Sawyers and others were seeking the hidden trick that some cases of CML seemed to be playing on the drug, others were trying to find routes around it. In particular, patients who had not had a cytogenetic response after being on the drug for a year were being offered higher doses, 800 milligrams as opposed to the usual 400 milligrams. That approach was sometimes sufficient to overcome resistance, but not always.

By sequencing DNA from the cells of patients resistant to Gleevec, the underlying mechanism gradually came to light. Both upfront and acquired resistance, Sawyers could see, were caused by the same phenomenon that led to CML in the first place: genetic mutations. "As we looked deeper, and as others looked deeper, it became clear that there were many different mutations that could cause resistance," said Sawyers. By 2002, Druker and other researchers who'd been investigating the problem at OHSU had found several dozen mutations occurring in addition to the original *bcr/abl* abnormality. Sometimes, patients with upfront resistance had mutations that were not present in patients who responded well to Gleevec. Patients who relapsed appeared to have acquired mutations during the course of their care. Genetic alterations that had not existed at the outset were present a few months into treatment. And, Sawyers showed, when certain mutations were induced in samples from CML patients, the cell-killing effects of the drug stopped. The connection between additional mutations and lack of response was just as direct as the connection between Bcr/Abl and response.

But knowing that mutations were present still didn't fully explain the problem. Why would a newly accrued mutation prevent kinase inhibition? What was going on in the cell? What did it look like?

Those questions raised one that had until then gone largely unvoiced. Was Gleevec really working as everyone thought it was? The drug had been designed to target the ATP binding site on the Bcr/Abl tyrosine kinase, the exact point where the phosphate adhered to the enzyme. But there was really no way to confirm that this binding was actually happening inside the cell because no one could see it. From the chemists at Ciba-Geigy to the clinical trial investigators, everyone had made educated guesses about what the kinase looked like. No one knew for sure.

Just as Sawyers was puzzling over these questions, a man named John Kuriyan, at the University of California–Berkeley, came up with the answer. The solution had come from x-ray crystallography, a technique that was not advanced enough to be useful when Jürg Zimmermann was sketching out the design of the experimental molecule using pencil and paper. Kuriyan had shot x-rays at crystals and watched how the resulting beams of light bent as they hit the kinase, each angle enabling him to build a picture of the molecule. The technique brought the Bcr/Abl kinase out of the dark room of the cell and into the light of a computer monitor.

That image was immediately indispensable. Kuriyan could show exactly where the molecule bound ATP and exactly how Gleevec interrupted that process. Sawyers regularly flew from Los Angeles to Berkeley, where he would don 3-D glasses and stare at Kuriyan's computer while he manipulated the image of the kinase, rotating it to reveal its every angle. For Sawyers, it was like traveling to the dark side of the moon and realizing that it held the secrets of the universe.

The images took him right back to where Zimmermann had started when he crafted the anti-kinase molecule: the shape of the enzyme. Now, armed with this vibrant image, the kinase looking like an enlarged kidney bean with swirling ribbons delineating its various atoms and chemical bonds, Sawyers could see why additional genetic mutations stopped Gleevec from working. The mutations changed the shape of the kinase. When that happened, the fit that Gleevec had to the ATP binding site was no longer snug. Bcr/Abl could escape its grip and resume its haywire activity.

That explained the resistance. Cancer cells accrue mutations over time, and that evolution gradually changes the structure of the cells. As a result, a component that starts out just a bit different from its corresponding part in a normal cell eventually becomes totally foreign. Sawyers, Druker, and others had also figured out that some CML patients have other mutations present from the start that prevent them from ever responding to Gleevec.

Sawyers was thrilled to understand resistance, but he knew that the next important question now had to be tackled. Could the science behind resistance lead to better drugs? The success of the first tyrosine kinase had depended entirely on the fact that it blocked Bcr/Abl and only Bcr/Abl. The shape of the kinase and the medicine were thoroughly specific, which is why the treatment worked. How could a drug target a kinase that was pulled out of shape by new mutations? What were the chances of creating a chemical with enough wiggle room to block an even-more-mutated kinase?

36

THE FIRST FIVE YEARS

Following its first approval, a strident pace of expansion brought Gleevec to an ever-widening pool of patients. The first approval, in May 2001, was for patients with CML who'd not responded to interferon or were ineligible for a bone marrow transplant. Novartis had opened numerous clinical trials after the first phase I study had revealed the potential power of the drug. As those subsequent studies proved successful, the drug tumbled forward with more and more FDA and international approvals for the treatment of other CML populations. As other malignancies fell prey to the new drug, approvals expanded beyond leukemia.

In November 2001, Gleevec was approved in the European Union and Japan for adult CML patients who'd not responded to interferon therapy or who were in the accelerated or blast crisis stages of the disease. In February 2002, Gleevec was approved for the treatment of patients with GIST who were not eligible for surgery or in whom the cancer had metastasized following a previous other treatment.

In May 2002, the results of the phase III IRIS study—in which newly diagnosed patients were randomized to receive either Gleevec or interferon plus ara-C as their first treatment—were broadcast for the first time. In responses, survival time, and side effects, Gleevec was consistently superior to interferon. Halfway through the trial, a group of statisticians conducting an interim review of the data insisted that all the patients who'd been randomized to interferon be switched to Gleevec; the advantage was already so unequivocally clear that it would have been unethical to continue giving patients the older medication. The subsequent publication in the *New England Journal of Medicine* in 2003 showed that after eighteen months, 75 percent of newly diagnosed patients had reached a complete cytogenetic response. The conditional approval of Gleevec for CML would soon become a standard approval. The drug would be widely available for newly diagnosed patients, doing away with the need to first try interferon or undergo a bone marrow transplant.

In December 2002, the drug was approved in Europe for children with CML who were not candidates for a bone marrow transplant. In May 2003, the FDA approved it for children in the early stages of the disease who either had a recurrence after a transplant or for whom interferon had stopped working. In 2006, Gleevec was approved for the treatment of acute lymphoblastic leukemia in cases where the leukemia is positive for the Philadelphia chromosome; a rare skin tumor called dermatofibrosarcoma protuberans; a group of precancerous blood conditions called myelodysplastic syndrome; aggressive systemic mastocytosis, a rare condition that strikes the connective tissue; and two linked blood conditions known as hypereosinophilic syndrome/chronic eosinophilic leukemia. In 2008, the GIST indication was expanded to include postsurgical treatment.

Eventually, Gleevec would be approved in 110 countries. The FDA approvals would span six different diseases. In addition to the growing roster of approvals (all under the name Glivec outside the United States), the continued health of CML patients taking the drug also escalated sales. CML became a peculiar phenomenon: It was the only cancer with an increasing number of survivors. As the population of people with CML continued to grow—more people diagnosed each year added to the number of patients who were no longer dying from the disease—so did sales of Gleevec. In 2006, the sale of Gleevec increased by 17 percent. By 2007, cancers other than CML accounted for 30 percent of the sales. In 2007, annual sales of Gleevec totaled more than $2.5 billion.

Novartis forged ahead into other rare-disease territory. In 2007, a *Forbes* article titled "Big Bucks," documenting the strategy, reported that the company had drugs in the pipeline for the treatment of at least six conditions diagnosed in fewer than 200,000 people in the United States per year, the criteria for a disease to be considered orphan. In 2009, Ilaris, the company's drug for a rare autoimmune disease driven by a single genetic mutation, was approved, with annual sales totaling about $26 million. Gleevec had proved the principle of targeting the underlying genetic cause of a disease, and now it had proved another: Rare diseases are a profitable market. (Novartis's effort to get Ilaris approved for the treatment of gout, which would have turned the drug into a true blockbuster—defined as annual sales of $1 billion or more—was rejected by the FDA in 2011.)

WITH EVERY PASSING year after the initial approval in 2001, the question of whether patients' responses would last persisted. The trials may have been completed, but the experiment continued. There was no track record, no backlog of evidence, no years of experience. At that point, the goal was to reach five years.

At the same time, a major leap had been made in measuring responses. A diagnostic test known as polymerase chain reaction had become widely available since the early clinical trials. The test enabled a deeply penetrating view into blood cells taken from CML patients. It was used to find how many cells contained the *bcr/abl* gene and was powerful enough to spot a single abnormal cell among 100,000 normal ones. Results were calculated in terms of log reductions, with each one signifying ten times fewer cells with the *bcr/abl* gene. Now, in addition to hematologic and cytogenetic, another response was added to the mix: molecular. If the polymerase chain reaction test showed a 3-log reduction over time, then the patient was considered to have a major molecular response. It was as close as a CML patient could come to being cured.

By 2006, the phase III IRIS trial had been ongoing for five years. Druker and the other investigators assembled everything they knew about the patients who'd been taking the drug all that time. Their report, published in the *New England Journal of Medicine,* prompted yet another collective sigh of relief. Of the 350 patients who'd had a complete cytogenetic response within a year of starting treatment, the disease had stopped progressing in nearly all. The IRIS trial had included polymerase chain reaction testing, and the results were startling. Every single patient who had a major molecular response within eighteen months of starting treatment was still alive at the five-year mark. Among the 382 patients who'd been taking Gleevec since the study opened in 2001, 340 were still alive in 2006. "It is currently recommended that imatinib therapy be continued indefinitely," the authors concluded.

37

THE SECOND GENERATION

Now that Sawyers knew that drug resistance was caused by changes in the shape of the binding site, he envisioned a solution. A drug that still adhered to the kinase but with a structure that allowed for those changes might still block the enzyme from binding ATP. "If you could find a drug that . . . was less demanding, more promiscuous, then it should work," Sawyers said. As he traveled around to various cancer meetings, he started testing the waters of this new concept, to see if it grabbed anyone's interest. Soon after one of those meetings, he got a call from a scientist at Bristol-Myers Squibb. When the scientist heard Sawyers speak about the problem, his mind had turned to a compound his lab created.

Called dasatinib, the compound had been created to inhibit T cells in the immune system. CML wasn't a target of investigation at the industry lab, but when the drug was screened against a group of kinases, the scientists there noticed that it blocked Abl. The company sent Sawyers a sample of the compound, and he tested it in a collection of cell samples from patients who were resistant to Gleevec. In nearly every sample, dasatinib blocked the kinase. With Gleevec having already charted the course for tyrosine kinase inhibition, the drug went quickly into clinical trials, where its anti-CML activity was confirmed.

As the investigation of dasatinib proceeded, exact patterns of Gleevec resistance emerged. Patients who started taking Gleevec very soon after being diagnosed with CML tended to continue doing well; the chance of relapse among this population was about 4 percent for the first five or six years. But many patients who started the drug several years after the diagnosis tended to stop responding to it at some point. The extra time had left a window for additional genetic mutations. The slow accrual of abnormalities went hand in hand with resistance.

Bristol-Myers Squibb may have been the first company to bring a second-generation tyrosine kinase inhibitor to clinical trials, but Novartis was not far behind. Ever since Gleevec had been approved, Alex Matter had been wondering if the chemists could do even better. Could the compound be made stronger? If so, would that improve patients' outcomes even more? Matter, Zimmermann, and the other researchers at Novartis were eager to find out.

Diving into a second-generation tyrosine kinase inhibitor stirred up yet another fight for the ever-combustible Matter. "You're going to destroy our franchise," the marketing team warned him. If word got out that Novartis was working on another compound, then patients and the public at large would infer that Gleevec was not the best possible medication. Matter insisted that if Novartis didn't pursue the next such inhibitor, then the competition would.

The marketing team had to relent. After all, they knew the patent would expire and exclusivity would end in a few years, leaving the CML market wide open. Novartis had to have another drug waiting in the wings to fill its place. Vasella was also well aware of the pressure to keep creating new drugs. In his view, that was the whole purpose of patent expiration. "Unless you innovate, you're dead," he says, "which is a pretty good system." Despite the fact that the Gleevec trials had just proved the principle of kinase inhibition, murmurs of skepticism about creating an even better inhibitor filled the company's hallways. "Most people said no, we have no chance to beat Gleevec," Vasella recalls. "But we did." In 2002, the next-generation tyrosine kinase inhibitor, called nilotinib, was synthesized. It was a direct descendant of imatinib but with better binding capacity.

In separate phase I clinical trials, dasatinib and nilotinib were given to groups of patients who had stopped responding to Gleevec. Just as with the first trial in 1998, dose-escalation studies were conducted for the sole purpose of finding the most effective dose at which the drugs could be given safely. But again, the data far exceeded that goal. Of twelve patients with chronic-stage CML, eleven had a complete hematologic remission with nilotinib. In the dasatinib study, that response was seen in thirty-seven of forty chronic CML patients. The drugs also worked in the more advanced stages of the disease. Patients with large numbers of blast cells in their marrow saw their blood counts return to normal. Cells carrying the mutant Philadelphia chromosome gene decreased in number. In 2006, the FDA approved dasatinib (the brand name is Sprycel) for the treatment of CML in patients for whom imatinib didn't work. Both of the second-generation drugs had side effects, but as with their predecessor, nothing that was intolerable. A year later, nilotinib (brand name Tasigna) was approved for the same indication.

With first- and second-generation tyrosine kinase inhibitors available for CML, the number of patients living with the disease continued to expand. "The prevalence, the number of people living with CML, keeps doubling" said Michael Mauro, the oncologist who'd joined Druker's team in 2000. Before Gleevec, there were 25,000 to 30,000 people living with CML in the United States. "By mid-century, it will be a quarter of a million," Mauro said. "Ten times greater [than before Gleevec]." That number will eventually plateau because the number of new annual diagnoses remains constant. But the expense of CML care will become even more formidable in the coming decades. "It is a good problem to have, but we have to be thoughtful about . . . the best way to manage a patient indefinitely."

The cost of treating CML for decades combined with the disappearance of the Philadelphia chromosome in many patients has led to speculation that some patients may be able to stop taking the drug after a prolonged period of time. That question is now being investigated, but many clinicians are skeptical and hesitant to take the risk. Some patients who had achieved a complete molecular remission—no detectable sign of the cancer, even with the most powerful disease-spotting equipment—began experimenting with stopping Gleevec after several years. In some of those patients, the cancer soon returned.

People who stop taking Gleevec run the risk of the drug not working as well when they resume, though the second-generation drugs would then be an option.

The long-term management of healthy patients wasn't the only problem remaining. One group of CML patients did not respond to any of these drugs. These patients had a mutation known as T315 that was impervious to all of the tyrosine kinase inhibitors at any dose. By the end of the first decade of the twenty-first century, patients found to have this mutation were routinely told that a bone marrow transplant was their best shot at survival. In 2009, however, a compound designed specifically for patients with this mutation, ponatinib, entered clinical trials.

Hans Loland is a CML patient from outside Seattle. After failing to respond to any drug treatment and watching his best friend, who also had CML, die after a bone marrow transplant, Loland was offered a spot in the ponatinib trial by Mauro, who was leading the trial at OHSU. Loland knew it was his last chance. Three months after starting on the drug, he had a complete cytogenetic response. Two years later and with a five-month-old son, Loland now has to stop himself from worrying that the response will disappear. "Before, I had everything to gain," he said. "Now, I have everything to lose."

In late 2012, ponatinib (brand name Iclusig) was approved by the FDA for treatment of CML. There are now approved tyrosine kinase inhibitors for all known permutations of the disease.

38

A GLEEVEC FOR EVERY CANCER

By 2011, the landscape of cancer research looked vastly different from when Druker started, and it seemed an alien planet compared to Nowell and Hungerford's early days. The advent of Gleevec had let loose an avalanche of change for cancer research and treatment.

Finding the driver mutations behind different types of cancer became a central thrust of research. "If you know it's broken, you fix it," is how Druker summarizes the pursuit. As the data on tyrosine kinase inhibition continued to mature, Druker turned his attention to tumor sequencing. By identifying the cluster of mutations that matter to a particular cancer, investigators could create treatment regimens tailored specifically to individual tumors. Currently in the thick of the search, Druker sees routine tumor sequencing as the near future for personalized medicine. Genome sequencing of an individual's entire DNA, with the aim of creating a personal profile of susceptibilities, prognoses, and prescriptions, is, he said, the fifty-year vision.

But tumor sequencing has been a steep mountain to climb. As researchers continued to profile cancer after cancer, reports were turning up as many as 200 mutations, and at the same time finding that only as many as five could be crucial to the disease. Among those five mutations, some may be easily drugged, and others not so much. At OHSU, Druker has continued to expand screening efforts to include more than 1,000 different mutations across 200 genes. Today, when intriguing mutations are identified, the storehouse of compounds accrued from years of academic and industry efforts can be screened for activity against them.

Alongside searching for new drugs is a quest to create a smoother road for bringing them to market. The eighteen-year odyssey of moving the first tyrosine kinase inhibitor from the labs of Ciba-Geigy to doctors' offices in the most remote US towns did more than change scientific thinking about cancer. It also transformed views on how drug development should best be done.

In the wake of Gleevec's approval, Druker and many other investigators began to think up new ways of testing experimental drugs in humans, "a completely reconfigured clinical trial process and FDA approval process." In the clinical trials for STI-571, the first fifty chronic-phase CML patients had the same response rate as the last fifty. Studying the drug on this homogenous population, he said, is akin to "running 500 identical twins through that study." Valuable as it was to have that much data, the same results would have been had if the study had stopped at fifty or a hundred enrollees.

Some researchers, including Druker, say that a conditional approval based on those fifty or so patients would work just as well as a conditional approval after a phase II study. Full approval could be withheld until safety data from the full-scale trial of 1,000 patients is cleared, but insurers would begin covering expenses for the remaining 900 patients, enabling drug companies to start recouping their investment sooner, greatly decreasing the cost of bringing new drugs to market.

Novartis and some other companies are already trying this approach as a way to prove concepts earlier in the development process. A small patient population with very similar disease characteristics is enough to show whether the rationale behind the design is correct. "Even if you have only twenty patients, if you then have a number who respond, you can say probably my hypothesis is right," said Vasella. Whether the FDA will consider small populations for conditional approvals, however, remains uncertain.

Druker and Vasella were also left with a similar outlook on whether, and how, academia and industry can work together. In the aftermath of the clinical trials and in light of the evolving direction of cancer research, Druker could see that improved collaboration would be essential to the future of drug development. "The drug companies aren't evil," said Druker. "They make drugs, and we should help them."

Federal funding has remained flat in recent years, the revenue for hospitals attached to academic institutions will likely not be rising anytime soon, and philanthropic support has also faced challenging times economically, which leaves industry an indispensable source of funding. Plus, drug companies exist to make drugs and are well equipped to do so. "[Gleevec] was a successful collaboration between academic and industry," Druker said. "Either side probably has some complaints about the other side, but it worked. We got a drug from a drug company to people, and it works, and that was an academic/industry collaboration." Vasella also began wishing for more open collaboration with academia. In light of that need, the increased vigilance of bias among physicians who work with pharmaceutical companies seems, to him, detrimental. "I think you have to work together and you have to trust each other," said Vasella. "And you can't see always the evil in things, [and be] so focused on believing in rules and regulations and disclosure."

Druker could see now how the Bayh-Dole Act made life unexpectedly complicated for people interested in such collaborations. Passed in 1980, the act, also known as the Patent and Trademark Law Amendments Act, gave universities ownership of the intellectual property created there, even when that creation was made possible by federal grants. That change in law—before, any invention made using federal grant money was public domain—allowed for the biotech revolution.

But an unintended consequence was runaway bureaucracy. Technology transfer offices sprang up in academic institutions where research labs were creating or testing substances that might advance medicine. The offices were there to manage the transfer agreements, but the process grew increasingly complex, making it difficult for academic investigators to work with companies. When Nick Lydon sent compounds to Druker to test in his lab, executing the contracts for that transfer took minutes. Nearly twenty years later, the process has become incredibly laborious, with academic institutions on hyperdrive about protecting any potentially lucrative additions to an experimental compound sent from a pharmaceutical company. "They're fixed in their view of getting the best deal out of every deal," said Druker, who feels that approach only wastes time. "I'm not willing to milk the last dollar out of every deal because it will just make it difficult to work with industry."

People involved in experimental cancer drug development also have become increasingly aware of the need to test combinations of experimental drugs in clinical trials. If cancers are driven by multiple mutations that can be targeted by multiple drugs, then being able to test two or more genetically targeted drugs at once is essential to advancing treatment. The problem is that the FDA generally does not allow two investigational drugs—meaning two drugs that have not been approved—to be tested in the same clinical trial. As Vasella continued to lead Novartis into creating drugs for new therapeutic areas, with an increased focus on rare diseases that has been extraordinarily successful for the company, he could see that these restrictions were holding up the pace of research. "One should be much more reasonable and flexible, and look at the rationale," said Vasella, "and then decide from case to case based upon a concrete request." At the very least, he said, companies should be able to collaborate or synthesize one another's drugs for animal studies.

Other pressing issues persist. In recent years, Michael Mauro, in addition to caring for leukemia patients at OHSU, mentored physicians in India, Africa, and Southeast Asia. Immediately Mauro was confronted with problems stemming directly from poverty and scant access to medical care. A physician in Africa told Mauro that even when he knows patients have CML, he often can't get them the proper medicine because the diagnostic test to confirm the presence of Bcr/Abl requires resources that patients and hospitals don't have. A couple of years ago, Mauro met a young woman from Mexico with two children. When she was diagnosed with CML, she was told she could have Gleevec for a while but that ultimately she would need a bone marrow transplant because a matching donor had been identified for her. Her insurer would not pay for continued medication and insisted she undergo an extremely dangerous procedure instead. "She knew [transplantation] was risky and would take her away from her family," Mauro said. "So she came into the US illegally." She found her way to OHSU, where she received emergency coverage and patient assistance from the drug manufacturer.

Novartis has also become embroiled in a prolonged patent fight in India, where generic Gleevec has been made despite the company's insistence that the drug is still patent protected. Vasella cheers the company's patient-assistance program for being generous internationally, not only in the United States. "In India we have thousands of people who get [Gleevec] for free, and we have about 2,000 who pay for it," he says. But the Indian government has accused the company of evergreening, extending the life of the patent by making ever-so-slight adjustments to the compound, altering it just enough to warrant patent extension without changing the underlying mechanism of the drug.

As that fight has worn on, Novartis has continued to raise the price of Gleevec in the United States. The current cost for a one-month supply is $6,328. From 2001 to 2011, sales of Gleevec worldwide totaled $27.8 billion. The US patent for Gleevec, number 5,521,184, originally set to expire on May 28, 2013, has been extended to January 4, 2015.

THE ADVENT OF Gleevec as a targeted drug has changed not only the course of cancer treatment, but medicine as a whole.

"The revolution in cancer research can be summed up in a single sentence: cancer is, in essence, a genetic disease," the eminent oncologist Bert Vogelstein is quoted in Siddhartha Mukherjee's Pulitzer Prize–winning book on the history of cancer research, *The Emperor of All Maladies*. Today, stories about tumor sequencing and genetic drivers of cancer are top news headlines daily. Tyrosine kinase inhibitor programs are commonplace at large pharmaceutical companies, as are new, small biotechs created to develop a single targeted compound. When drug developers imagine the best-case scenario for their rationally designed drugs, they are imagining Gleevec for CML.

In the decade-plus since Gleevec was approved, tyrosine kinase inhibitors have become a mainstay of cancer care. More than fifteen such drugs are now available. Among them are erlotinib for lung cancer, lapatinib for breast cancer, and sunitinib for kidney cancer as well as for GIST that hasn't responded to imatinib. All were approved in the past decade, and all have improved the odds for cancer patients, extending survival time and offering a less harsh treatment option compared to chemotherapy alone, although many of these new drugs are given in combination with traditional chemotherapy. Tyrosine kinase inhibitors are a $15 billion-a-year market, an amount that is projected to double over the next ten to fifteen years. The fact that these drugs can often be taken at home adds to the way they have transformed cancer care. For many patients, treatment can be bent around their lives, rather than the other way around.

The pharmaceutical industry has seized on the promise of tyrosine kinase inhibitors. Every large company has a kinase inhibitor pipeline: Novartis, AstraZeneca, Bayer, Johnson & Johnson, Merck, Pfizer, Eli Lilly, GlaxoSmithKline, and AstraZeneca, to name but a few. The number of potential kinase targets has become a vast landscape of technical research and endless abbreviations: PI3K, MEK, JAK1, JAK2, cyclic-dependent kinase inhibitors, the CAMK family, the TKL family, p38—again, to name but a few. All of these proteins are involved in one cancer or another, suspects in the search for its cause. All are encoded by genes that are, in some way, abnormal, proto-oncogenes that have transformed into oncogenes, just as *abl* is when it is translocated next to *bcr,* just as the *src* gene was altered when it was integrated into the Rous sarcoma virus. Today, more than 500 kinase inhibitors are in development at more than 250 different companies, targeting more than 200 different proteins.

Targeted therapy, the catchier name for rational drug design, extends far outside the realms of the kinase. The approaches divide into two main types: small molecule inhibitors that fit inside the cell, and monoclonal antibodies, which don't. The success of Herceptin for breast cancer, approved just as the phase I trial of STI-571 was reaching an effective dose level, gave rise to a growing number of monoclonal antibodies. These drugs block targets outside the cell or on its surface using the body's natural immune system. Among the small-molecule inhibitors, tactics beyond kinase inhibition include inducing cancer cell death, delivering radioactivity to cells containing cancer-causing molecules, cutting off the blood supply to tumors, and binding a structure called the proteasome as a way to kill cancer cells. Scientists are on a continual search for new pathways to target. Owen Witte, still researching in California, is currently trying to remove lymphocytes, part of the immune system, from the body, genetically engineer them to kill cancer, and then reinject them into the body to accomplish that goal.

So far, targeted therapy has hardly lived up to the expectations set when Gleevec was approved. The benefits have, for the most part, been incremental. Patients' lives are extended by months, a significant and important amount of time, but hardly the normal life span that people living with CML—many of whom don't think of themselves as patients anymore—are now experiencing.

The failure of tyrosine kinase inhibition or other targeted therapies to transform other cancers into tolerable chronic conditions has generated skepticism about the future of the approach. Some scientists point to genetic instability—the constant and unpredictable appearance of new mutations—as a guarantee that cancer cells will eventually become resistant to whatever therapy is thrown their way. "Therapies for Cancer Bring Hope and Failure," ran the headline of a 2010 article in *The New York Times* by Andrew Pollack providing a where-we-are-now appraisal of the targeted-drug approach. "We've gone through a very rapid period of high expectations, maturation, and disappointments," Dr. J. Leonard Lichtenfeld, deputy chief medical officer of the American Cancer Society, was quoted as saying in that article. "I think there was almost a naiveté that if we could find the target, we would have the cure." Many targeted therapies have come on the market in the past ten or so years, and many are disappointments.

The evolving survival data from the initial Gleevec trials continues to be closely scrutinized, in part to understand how the drug works and in part to inform public and scientific opinion on targeted therapy in general. In 2012, ten-year survival data on 368 patients treated at MD Anderson Cancer Center were published. All of the patients were from the earliest studies, restricted to those who had stopped responding to interferon. Ten years after entering the clinical trial, 68 percent—250 patients—were still alive. CML was progressing in less than five of those patients. A portion of the surviving patients had stopped responding to Gleevec at some point over the years, and had then been switched to a second-generation inhibitor. Almost all of the patients who'd survived ten years since enrolling in the study had had a complete cytogenetic response—no evidence of the Philadelphia chromosome—emphasizing the strong link between this depth of response and survival. It is important to note that this particular study wasn't a comparison; no other treatment arm was tracked simultaneously, a design that tends to diminish the robustness of the findings, as does the fact that the study was done at just a single cancer center. But the ten-year survival data can be viewed in light of the pre-Gleevec median survival time of four to six years. Before 2001, no one diagnosed with CML was told he or she might live for ten years or more.

To Druker, high expectations were part of what will come to be seen as just the very earliest days of a new time for cancer care. As he sees it, the clarity of knowing that cancer is a genetic disease is the starting point for what may require twenty years of research or more before the next Gleevec comes along. As tumors are genetically sequenced, cancer patients will be diagnosed according to their molecular profile. At the same time, ongoing research to profile experimental compounds according to what molecules they likely target can dovetail with tumor profiling. "Let's sequence every tumor and figure out what drugs match that tumor," said Druker. Genetic profiling of patients can also inform treatment decisions, because some patients may metabolize drugs faster than others, for example, a quirk that could be spotted in a DNA sequence. Then, patients can be matched to a treatment for their particular tumor and personal genetic profile.

The approach has been widely chronicled, hailed as the next big thing for cancer, and greeted with equal parts skepticism and optimism. Michael Mauro cites the recently approved drug crizotinib, for lung cancer, as the latest proof of the principle. The drug targets an abnormality present in just a subset of lung cancer patients and will be given only to those patients. A mutation in a gene called *b-raf* was recently found to be present in all patients with hairy cell leukemia. Although the disease can't be triggered by forcing a *b-raf* mutation, it can be stopped by blocking the mutation. Hairy cell leukemia is rare and already has an effective treatment, but Mauro sees the finding as evidence of the validity of continuing to uncover the genetic roots of cancer. "It's history starting to repeat itself," he said.

In addition to his work at OHSU, Druker has also launched Blueprint Medicine, based in Cambridge, Massachusetts, with Nick Lydon. The company's research is aimed at screening tumors and compounds for potentially relevant genetic abnormalities. Other companies are whittling down the time required for genome sequencing to forty-eight hours. Others are focused on developing technology to identify the four or five driver mutations out of the 200 or more that may be present in a given tumor sequence at any given moment.

"There would be a Gleevec for every cancer," Druker said of his hope for the future of cancer treatment, the drug now as much a symbol as it is a medication. The drug continues to guide his vision for cancer research, one that is inextricably linked to his experience over the last twenty years. The lessons of Gleevec—the biologic principle that it proved and the business model it created—continue to inform his work. "It's all about the target. Identifying the right target, getting a good drug for that target," said Druker, "and not worrying about what the market size is."

Epilogue

SURVIVAL TIME

Gary Eichner, Druker's patient that chilly February morning in 2012, knew none of this history as he lay on the exam table with a needle stuck into his lower back. He had never heard of Peter Nowell or David Hungerford. He knew nothing about the Abelson virus and how researchers had used it to unravel the cause of CML. He was unaware of the legacy of kinase research, of the painful treatments that he'd never have to experience, of the role that his own doctor had played in securing his future. And as Carolyn Blasdel told Eichner the good news over the phone, that none of his white blood cells contained the Philadelphia chromosome, he had no idea of just how much history was packed into that single sentence—words that were, for him, the difference between life and death.

What he did know, and perhaps all he needed to know, was that he would live. That he would be there for his teenage son for the foreseeable future. That a disease that might have killed him would instead probably have little to no impact on his life and the lives of his loved ones. "What is my chance in the next five years of not relapsing? It looks good," he said. "I wake up every morning feeling extremely fortunate."

THE FINAL ISSUE of the *STI Gazette* was published in the spring of 2002. With Gleevec now approved and widely available, and the tight-knit trial community now dispersed as people resumed their normal lives, the newsletter had run its course. Orem was also ready to move on, though her home was now firmly in Portland. She and her husband had bought a house there, and she'd taken all her belongings out of storage. It was time to lay back down some roots. "I think I'm going to live," she'd said to her husband. "I don't think we should be thinking about this as a temporary thing." She would continue to do well on the drug, remaining active with her family, including the grandchildren who'd been born after she was diagnosed. By 2012, she held the record for the longest amount of time on Gleevec.

Others she'd known over the years were moving on, too. Suzan McNamara would continue to do well. She married the man who'd traveled with her to Portland when she'd been at her sickest. After several years of agreeing to requests from Novartis to speak about Gleevec and even appearing in some ads, McNamara retreated from the role of spokesperson. She didn't want to think of herself as a CML patient anymore; she just wanted to live her life.

Bud Romine never suffered from CML after the Gleevec trials. He died years later of unrelated causes.

Closing the newsletter was a bittersweet farewell for Orem. As in nearly every issue, Druker contributed his personal message. "Those of you who were the early participants in these clinical trials are the true pioneers. I will continue to greatly value our partnership in bringing Gleevec to the forefront. Patients have been extraordinarily generous in their thanks. No award, no media appearance, and no titles will replace the feeling of knowing that my work has made a difference in people's lives," he wrote.

Michael Mauro wrote likewise. "The best way to sum up these efforts is to say thank you, to you, the heroes, the brave souls—who were pioneers in historical trials that redefined treatment of an illness."

After some final brief memoirs, Orem signed off, her final words capturing exactly how the drug had changed her life and the lives of so many others.

"Thanks," she wrote. "Enjoy life."

GLOSSARY

ABELSON VIRUS A cancer-causing virus discovered by Herb Abelson. While researching the cellular mechanisms of cancer, he exposed healthy mice to the Moloney virus, which captured the *abl* gene from their DNA to become the Abelson virus. The virus induces B-cell tumors in mice.

abl The gene that encodes the Abl protein kinase. When combined with the gene *bcr* via a translocation, it forms *bcr/abl*, the mutant gene that induces CML.

ABL The protein product of the *abl* gene, which is named for the Abelson virus. The fusion protein Gag/Abl, a tyrosine kinase, drives the cancer-causing action of the virus. Another tyrosine kinase, Bcr/Abl, is part of the mechanism that causes CML.

ACUTE LEUKEMIA A type of leukemia characterized by rampant production of nonfunctioning blood cells. This type of cancer progresses rapidly, whereas chronic leukemia progresses slowly. Acute leukemias are the most common types of cancer in children.

AMINO ACIDS Organic molecules that serve as the building blocks of all proteins.

ANTIBODY A protein produced by the immune system to target harmful intruders such as viruses. In biological research, antibodies are often a useful way of detecting a substance; if an antibody is present, its antigen, or target, must be present.

ANTIGEN The target of an antibody. Each antibody targets a unique antigen.

ASH The American Society of Hematology. It was at the ASH annual meeting in 1999 that Brian Druker gave a landmark presentation on the effectiveness of STI-571, the drug that would become Gleevec.

ATP Adenosine-5'-triphosphate, an essential molecule used to store and transfer energy in cell metabolism. The normal function of a kinase is to take one phosphate from an ATP molecule and attach it to another protein in a reaction called phosphorylation. Uncontrolled phosphorylation of the protein responsible for white blood cell production causes CML.

B CELLS A type of lymphocyte, one of the varieties of white blood cells. B cells are activated by helper T cells and produce antibodies to fight invaders. B cells are also the target of the Abelson virus, a key tool in early cancer research.

bcr A proto-oncogene located on chromosome 22 at the point where the chromosome breaks and swaps genetic material with chromosome 9 (the "breakpoint cluster region") to become the Philadelphia chromosome. This translocation brings *bcr* into proximity with *abl*, to form the *bcr/abl* fusion gene that is the cause of CML.

BCR The protein product of the *bcr* gene, and a component of the fusion protein Bcr/Abl.

BCR/ABL The protein product of the *bcr/abl* gene, this mutant tyrosine kinase is the mechanism for CML. Bcr/Abl phosphorylates a protein that triggers the creation of white blood cells, driving this usually tightly regulated process out of control and resulting in excessive production of blast cells.

BLAST CELLS Immature white blood cells. CML is characterized by unregulated production of these nonfunctional cells, and the disease's progress can be measured by the concentration of blast cells in a patient's blood.

BLAST CRISIS STAGE The final stage of CML, when the patient's blood consists of at least 30 percent blast cells. At this stage, without treatment, the disease progresses rapidly and survival time is limited.

CHROMOSOME An organized structure of DNA found in the nucleus of a cell. The human genome consists of 46 chromosomes, with 23 coming from each parent.

CHRONIC LEUKEMIA A type of cancer in which abnormal, poorly functioning blood cells are produced in excess. Onset is typically slow, and the disease generally affects adults.

CLINICAL TRIAL A series of studies in which a new drug is tested for effectiveness and safety. Several rigorous stages of testing are required before a drug can be approved by the FDA for marketing.

CLL Chronic lymphocytic leukemia, a cancer similar to CML that begins in the bone marrow but then moves to the lymph cells.

CML Chronic myeloid leukemia, the cancer caused by the Philadelphia chromosome mutation and targeted by the drug Gleevec. Before Gleevec, the most effective CML treatment could prolong life for only a few years, but the disease is now a manageable condition when treated with Gleevec or other tyrosine kinase inhibitors.

CYTOGENETICS The study of the connection between genes and diseases. Among patients with CML, a cytogenetic response to treatment means that the number of cells in the bone marrow containing the Philadelphia chromosome mutation, the root cause of the disease, are reduced.

DASATINIB A drug originally created to inhibit T cells in the immune system, it was found to be effective against CML and is part of the second generation of tyrosine kinase inhibitors for CML.

DNA The double-helix organizational structure of the genetic code. DNA is made of genes that encode the proteins responsible for nearly every biological function.

EGFR Epidermal growth factor receptor, a kinase that is a member of the erbB family of proteins. It has been considered a promising target for kinase inhibition because it is found in a variety of common cancers. Another member of the erbB family, Her2, is found in some breast cancers.

ENZYME A type of protein that facilitates various cellular processes. Protein kinases, which in a mutated form can cause cancer, are a type of enzyme.

FDA The Food and Drug Adminstration, the US organization responsible for approving new drugs for marketing by pharmaceutical companies.

FISH Fluorescence *in situ* hybridization, a method of analyzing DNA under a fluorescence microscope that highlights particular genes in different colors. This technique allowed scientists to investigate the particular mutation responsible for the Philadelphia chromosome and CML.

FOCUS ASSAY The technique of exposing healthy cells to a cancer-causing agent in a Petri dish and allowing the resulting cancerous cells to multiply. A method for investigating and quantifying external factors that can induce cancer.

gag A gene present in the Moloney virus, the progenitor of the Abelson virus. This proto-oncogene has the potential to join with *abl* and produce the harmful tyrosine kinase Gag/Abl.

GAG The protein product of the *gag* gene. A component of the fusion protein Gag/Abl.

GAG/ABL The protein product of the *gag/abl* oncogene. Gag/Abl, a tyrosine kinase, is responsible for the cancer-causing activity of the Abelson virus.

GENE A sequence of nucleotides on a strand of DNA or RNA. Genes are the basis of heredity and encode the proteins that drive most cellular functions.

GIST Gastrointestinal stromal tumors. A previously untreatable form of cancer, this disease has been found to respond to treatment with Gleevec.

GLEEVEC/STI-571 A targeted therapy for CML and the first kinase inhibitor to effectively fight cancer. Gleevec works by inhibiting the unregulated activity of the Bcr/Abl tyrosine kinase. Gleevec has also been found effective against other conditions such as GIST.

HEMATOLOGIC Relating to the study of the blood. Treatment for CML can be evaluated by measuring a patient's hematological response, ideally a reduction in their count of cancerous white blood cells.

IND Investigational new drug. A type of application seeking FDA approval for an experimental compound.

INTERFERON This drug was the only treatment for CML before Gleevec, but its effectiveness was limited and the side effects for patients were severe.

IRIS *I*nternational *R*andomized Study of *I*nterferon and *S*TI-571. The phase III trial of the drug that would later be named Gleevec. This final trial for the drug proved it to be an effective treatment for CML.

KARYOTYPE The number and type of chromosomes possessed by an organism. Deviations from a normal karyotype can indicate harmful genetic mutations.

KINASE A type of enzyme that is instrumental to starting processes in the cell. The term is derived from the Greek word *kinetic*, meaning "motion." Kinases set off a cascade of signals by plucking a single phosphate from a molecule of ATP and placing it on a protein, activating that protein to do its job. Mutated and malfunctioning kinases are the cause of many cancers.

KINASE INHIBITOR A compound that blocks the action of a kinase. Gleevec is a kinase inhibitor that binds to the kinase Bcr/Abl at the site where Bcr/Abl normally grabs onto ATP. This arrests the progress of CML because, without ATP, Brc/Abl can no longer phosphorylate the protein responsible for blast cell production.

KIT A tyrosine kinase is inhibited by Gleevec. Kit plays a role in some GIST cancers, and this involvement led to the use of this drug to treat GIST.

LEUKEMIA Cancer of the blood or bone marrow. Many early cancer researchers focused on these and other "liquid cancers," as the cancer's presence in the bloodstream made quantifying its progress much more straightforward than with solid tumors.

MOLONEY VIRUS An RNA virus that causes cancer in mice. Herb Abelson's research on the virus led to the discovery of the Abelson virus, a key tool in the exploration of cancer.

MUTATION An alteration to the nucleotide sequence of an organism's DNA. While many mutations have no effect at all, some bring about profound changes. The Philadelphia chromosome is the result of a type of mutation known as a translocation.

NIH The National Institutes of Health, a US government agency under the umbrella of the Department of Health and Human Services. The primary outlet by which the US government funds scientific research relating to medicine.

NUCLEOTIDES The building blocks of DNA and RNA. The four molecules adenine, guanine, cytosine, and thymine are the individual units that make up genes in DNA. In RNA, uracil takes the place of thymine.

ODA The Orphan Drug Act, a US law passed in 1983 meant to encourage the development of drugs for relatively rare diseases. The law includes financial incentives for drug developers, such as federal funding for clinical trials, tax benefits, and a guaranteed period in which they are the exclusive seller of the drug.

OHSU Oregon Health and Science University. Brian Druker began investigating a new treatment for CML at this university in 1993, and in subsequent years it became known as a premier facility for the treatment of leukemia and other cancers, largely due to Druker's work.

ONCOGENES Genes that cause cancer. Research has revealed that many oncogenes are mutations of genes normally found in healthy organisms that, in their mutated form, can trigger cancer.

PDGFR Platelet-derived growth factor receptor, a kinase found in excess in many types of cancer. Like PKC and EGFR, PDGFR was researched as a target for kinase inhibition because of its presence in several common cancers.

PHASE I TRIAL The first stage of human drug testing, a trial that enrolls a small number of patients and begins with very small doses of the drug, gradually increasing the dose over time and monitoring the results. The main goal is to ensure that the drug is safe.

PHASE II TRIAL The second phase of human drug testing, a larger trial that involves hundreds of patients at multiple locations. This phase of testing begins to investigate the drug's effectiveness.

PHASE III TRIAL A large drug trial in which patients are randomly assigned either the experimental treatment being tested or the best existing treatment. The effectiveness of STI-571, or Gleevec, was tested against that of interferon.

PHILADELPHIA CHROMOSOME (PH, PH¹) The name coined to describe the abnormally short chromsome 22 found in patients with CML. The mutation behind the shortened chromosome was originally considered by many researchers to be a deletion, but was later found to be a translocation in which genetic material is swapped between chromosomes 9 and 22.

PHOSPHATE A molecule built of one phosphorus atom and four oxygen atoms. ATP, the "fuel" of living cells, includes three phosphate groups. A kinase enzyme initiates a chain reaction (called a signaling cascade or signaling pathway) by removing one phosphate group from ATP and placing it on another protein.

PHOSPHORYLATION The process by which a kinase places a phosphate onto another protein, beginning a chain reaction in a cell.

PKC Protein kinase C, a kinase associated with some common cancers. Like PDGFR and EGFR, PKC was originally thought to be a promising target for kinase inhibition.

POLYMERASE CHAIN REACTION A diagnostic test used to gauge a CML patient's molecular response to treatment. The test measures how many of the patient's blood cells are carrying the Philadelphia chromosome on a logarithmic scale.

POLYOMAVIRUS A cancer-causing virus used in early studies of the mechanics of cancer. Research into polyomavirus revealed that some kinases, particularly tyrosine kinases, are involved in cancer.

PROOF OF PRINCIPLE A demonstration that a new drug actually works as expected. While many experimental drugs work by a similar mechanism to a drug already on the market and therefore rest on a proven principle, experimental drugs that work by a new, hypothetical mechanism can only gain proof of principle through testing.

PROTEIN A relatively large organic molecule made of amino acids. Proteins are instrumental to many vital processes that take place in a living cell. The amino acid sequence of a protein is encoded by an organism's DNA, so proteins can be thought of as the means by which an organism's genes guide the physical processes in its cells.

PROTO-ONCOGENE A normal gene that has the potential to become a cancer-causing oncogene. For example, *bcr* and *abl* are normal genes that, when combined via the Philadelphia chromosome mutation, form the oncogene *bcr/abl* which leads to unrestrained production of white blood cells.

RECOMBINANT DNA A process in which strands of DNA are cut and "pasted" together using specialized, targeted enzymes. Among other capabilities, this technology allows scientists to separate specific genes from the larger genome and study them individually.

RED BLOOD CELLS Specialized cells that deliver oxygen to other cells in the body.

RETROVIRUS A virus made of RNA, rather than DNA. These viruses contain the enzyme reverse transcriptase, which allows the virus to make DNA out of RNA as part of its replication process, instead of the other way around. DNA from a retrovirus may then become integrated into the host cell's own genome.

REVERSE TRANSCRIPTASE The enzyme that "reads" the nucleotide sequence of RNA and builds the corresponding sequence of DNA.

RNA The key intermediary between DNA and its protein products. DNA in the nucleus is translated into RNA, which then leaves the nucleus and is translated into the appropriate proteins in the cell. Retroviruses contain RNA rather than DNA.

ROUS SARCOMA VIRUS (RSV) This virus, discovered in the early 20th century, provided the first evidence that cancer could be triggered in cells by an infection. Later, the new technique called a focus assay revealed this virus to be a retrovirus.

SRC The protein product of the *src* gene. Src is a tyrosine kinase found in normal cells, but has the potential to cause cancer when expressed by a mutated oncogene.

src The gene responsible for the cancer-causing effect of the Rous sarcoma virus. Like *bcr* and *abl*, *src* is a proto-oncogene that is normally found in healthy cells but becomes cancerous when mutated.

STAUROSPORINE One of the first compounds discovered to be a kinase inhibitor (specifically targeting the kinase PKC). Staurosporine is an antifungal agent produced naturally by bacteria.

T CELL A type of lymphocyte, one of the varieties of white blood cell. Most T cells are either "killers" that target foreign substances, or "helpers" that trigger other killer cells to act.

TARGETED THERAPY A treatment that works by targeting a specific molecule or chemical in the body. Gleevec is a form of targeted therapy, aimed at the kinase Bcr/Abl. Another example is tamoxifen, which blocks estrogen and is used to treat some types of breast cancer.

TRANSLOCATION A genetic mutation in which two chromosomes exchange genetic material. The Philadelphia chromosome involves a translocation between chromosomes 9 and 22.

VIRUS A microscopic biological agent that can only reproduce by infecting living cells. Most viruses consist of just an external protein shell containing a short sequence of DNA (or RNA). A virus infects a host cell and uses the host's own replication machinery to reproduce.

WHITE BLOOD CELLS Immune system cells that defend the body from diseases. CML, like other leukemias, causes excessive production of defective white blood cells.

| REFERENCES |

INTERVIEWS

Interviews with academic scientists, doctors, industry researchers, patients, and industry executives served as a primary source for this book. The following individuals, listed in alphabetical order, were interviewed by the author. Some interviews were conducted in 2007; most were conducted in 2012.

Herbert T. Abelson, David Baltimore, J. Michael Bishop, Carolyn Blasdel, Clara Bloomfield, Sarah Bowden, Elisabeth Buchdunger, Renaud Capdeville, Sir Philip Cohen, Joel Crouch (by e-mail), George Daley (by e-mail), Brian Druker, Gary Eichner, Ray Erikson, Emil Freireich, Jennifer Gangloff, Steven Goff, John Goodman, Alexandra Hardy, Brian Hemmings, Alice Hungerford, Tony Hunter, Kara Johnson, Helen Lawce, Hans Loland, Nick Lydon, Alex Matter, Suzan McNamara, Kelly Mitchell, Peter Nowell, Frank Orem, Judy Orem, Beverly Alex Owen, Peter Parker, Naomi Rosenberg, Janet Rowley, Charles Sawyers, Moshe Talpaz, Peter Traxler, Daniel Vasella, Jackie Whang-Peng, Owen Witte, and Jürg Zimmermann.

ORIGINAL RESEARCH

Insights, history, data, and other information were culled from the scientific literature, conference presentations, newspaper and magazine articles, and a host of other works available in print or online. The following peer-reviewed papers, typically referred to as "original research" among the scientific community, are the published results of the most seminal laboratory experiments and clinical trials chronicled in this book, from the first observation of the Philadelphia chromosome, reported in 1960, to the most recent survival figures among CML patients who have been taking Gleevec since 1998.

Ben-Neriah, Y., G. Q. Daley, A. M. Mes-Masson, O. N. Witte, and D. Baltimore. The chronic myelogenous leukemia-specific P210 protein is the product of the bcr/abl hybrid gene. *Science* 233 (1986): 212–214.

Buchdunger, E., A. Matter, and B. J. Druker. Bcr-Abl inhibition as a modality of CML therapeutics. *Biochimica et Biophysica Acta* 1551 (2001): M11–18.

Buchdunger, E., J. Zimmermann, and H. Mett et al. Inhibition of the Abl protein-tyrosine kinase in vitro and in vivo by a 2-phenylaminopyrimidine derivative. *Cancer Research* 56 (1996): 100–104.

———. Selective inhibition of the platelet-derived growth factor signal transduction pathway by a protein-tyrosine kinase inhibitor of the 2-phenylaminopyrimidine class. *Proceedings of the National Academy of Sciences of the United States of America* 92 (1995): 2258–2262.

Carroll, M., S. Ohno-Jones, and S. Tamura et al. CGP 57148B, a tyrosine kinase inhibitor, inhibits the growth of cells expressing BCR-ABL, TEL-ABL and TEL-PDGFR fusion proteins. *Blood* 90 (1997): 4947–4952.

Daley, G. Q., R. A. Van Etten, and D. Baltimore. Induction of chronic myelogenous leukemia in mice by the P210bcr/abl gene of the Philadelphia chromosome. *Science* 247 (1990): 824–830.

De Klein, A., A. G. van Kessel, and G. Grosveld et al. A cellular oncogene is translocated to the Philadelphia chromosome in chronic myelocytic leukemia. *Nature* 243 (1973): 290–293.

Deininger, M. W., J. M. Goldman, N. Lydon, and J. V. Melo. The tyrosine kinase inhibitor CGP57148B selectively inhibits the growth of BCR-ABL-positive cells. *Blood* 90 (1997): 3691–3698.

Demetri, G. D., M. von Mehren, and C. D. Blanke et al. Efficacy and safety of imatinib mesylate in advanced gastrointestinal stromal tumors. *New England Journal of Medicine* 347 (2002): 472–480.

Druker, B. J., F. Guilhot, and S. G. O'Brien et al.; IRIS investigators. Five-year follow-up of patients receiving imatinib for chronic myeloid leukemia. *New England Journal of Medicine* 355 (2006): 2408–2417.

Druker, B. J., C. L. Sawyers, H. Kantarjian, D. J. Resta, S. F. Reese, J. M. Ford, R. Capdeville, and M. Talpaz. Activity of a specific inhibitor of the

BCR-ABL tyrosine kinase in the blast crisis of chronic myeloid leukemia and acute lymphoblastic leukemia with the Philadelphia chromosome. *New England Journal of Medicine* 344 (2001): 1038–1042.

Druker B. J., C. L. Sawyers, M. Talpaz, D. J. Resta, B. Peng, and J. M. Ford. Phase I trial of a specific ABL tyrosine kinase inhibitor, "CGP-57148B," in interferon-refractory chronic myelogenous leukemia patients. Poster presented at the Annual Meeting of the American Society of Hematology, 1998.

Druker, B. J., M. Talpaz, D. J. Resta, B. Peng, E. Buchdunger, J. M. Ford, N. B. Lydon, H. Kantarjian, R. Capdeville, S. Ohno-Jones, and C. L. Sawyers. Efficacy and safety of a specific inhibitor of the BCR-ABL tyrosine kinase in chronic myeloid leukemia. *New England Journal of Medicine* 344 (2001): 1031–1037.

Druker, B. J., M. Talpaz, R. J. Resta, B. Peng, E. Buchdunger, J. M. Ford, and C. L. Sawyers. Clinical efficacy and safety of an Abl-specific tyrosine kinase inhibitor as targeted therapy for chronic myelogenous leukemia. Plenary Presentation for the Annual Meeting of the American Society of Hematology, 1999.

Druker, B. J., S. Tamur, and E. Buchdunger et al. Effects of a selective inhibitor of the Abl tyrosine kinase on the growth of Bcr-Abl positive cells. *Nature Medicine* 2 (1996): 561–566.

Druker, B. J., S. Tamura, E. Buchdunger, S. Ohno, G. C. Bagby, and N. B. Lydon. Preclinical evaluation of a selective inhibitor of the ABL tyrosine kinase as a therapeutic agent for chronic myelogenous leukemia. *Blood* 86, supplement 1 (1995): 601a.

Eckhart, W., M. A. Hutchinson, and T. Hunter. An activity phosphorylating tyrosine in polyoma T antigen immunoprecipitates. *Cell* 18 (1979): 925–933.

Foulkes, J. H., M. Chow, C. Gorka, A. J. Frackelton, and D. Baltimore. Purification and characterization of a protein-tyrosine kinase encoded by the Abelson murine leukemia virus. The *Journal of Biological Chemistry* 260 (1985): 8070–8077.

Gale, R. P., and E. Canaani. An 8-kilobase abl RNA transcript in chronic myelogenous leukemia. *Proceedings of the National Academy of Sciences of the United States of America* 81 (1984): 5648–5652.

Goff, S. P., E. Gilboa, E. N. Witte, and D. Baltimore. Structure of the Abelson murine leukemia virus genome and the homologous cellular gene: Studies with cloned viral DNA. *Cell* 22 (1980): 777–785.

Gorre, M. E., M. Mohammed, K. Ellwood, N. Hsu, R. Paquette, P. N. Rao, and C. L. Sawyers. Clinical resistance to STI-571 cancer therapy caused by BCR-ABL gene mutation or amplification. *Science* 293 (2001): 876–880.

Groffen, J., J. R. Stephenson, and N. Heisterkamp et al. Philadelphia chromosomal breakpoints are clustered within a limited region, bcr, on chromosome 22. *Cell* 36 (1984): 93–99.

Heisterkamp, N., K. Stam, J. Groffen, A. De Klein, and G. Grosveld. Structural organization of the *bcr* gene and its role in the Ph' translocation. *Nature* 315 (1985): 758–761.

Heisterkamp, N., J. R. Stephenson, and J. Groffen et al. Localization of the c-*abl* oncogene adjacent to a translocation break point in chronic myelocytic leukemia. *Nature* 306 (1983): 239–242.

Hidaka, H., M. Inagaki, S. Kawamoto, and Y. Sasaki. Isoquinolinesulfonamides, novel and potent inhibitors of cyclic nucleotide dependent protein kinase and protein kinase C. *Biochemistry* 23 (1984): 5036–5041.

Hughes, T. P., A. Hochhaus, S. Branford, M. C. Müller, J. S. Kaeda, L. Foroni, B. J. Druker, E. Guilhot, R. A. Larson, S. G. O'Brien, M. S. Rudoltz, M. Mone, E. Wehrle, V. Modur, J. M. Goldman, and J. P. Radich. Long-term prognostic significance of early molecular response to imatinib in newly diagnosed chronic myeloid leukemia: An analysis from the International Randomized Study of Interferon and STI571 (IRIS). *Blood* 116 (2010): 3758–3765.

Iba, H., T. Takeya, F. R. Cross, T. Hanafusa, and H. Hanafusa. Rous sarcoma virus variants that carry the cellular src gene instead of the viral src gene cannot transform chicken embryo fibroblasts. *Proceedings of the National Academy of Sciences of the United States of America* 81 (1984): 4424–4428.

Kantarjian, H., S. O'Brien, G. Garcia-Manero, S. Faderl, F. Ravandi, E. Jabbour, J. Shan, and J. Cortes. Very long-term follow-up results of imatinib mesylate therapy in chronic phase chronic myeloid leukemia after failure of interferon alpha therapy. *Cancer* 118 (2012): 3116–3122.

Kantarjian, H., C. Sawyers, and A. Hochhaus et al.; International STI571 CML Study Group. Hematologic and cytogenetic responses to imatinib mesylate in chronic myelogenous leukemia. *New England Journal of Medicine* 346 (2002): 645–652.

Konopka, J. B., S. M. Watanabe, J. W. Singer, S. J. Collins, and O. N. Witte. Cell lines and clinical isolates derived from Ph1-positive chronic myelogenous leukemia patients express c-*abl* proteins with a common structural alteration. *Proceedings of the National Academy of Sciences of the United States of America* 82 (1985): 1810–1814.

Konopka, J. B., S. M. Watanabe, and O. N. Witte. An alteration of the human c-*abl* protein in K562 leukemia cells unmasked associate tyrosine kinase activity. *Cell* 37 (1984): 1035 1042.

Lydon, N. B., B. Adams, J. F. Poschet, A. Gutzwiller, and A. Matter. An *E. coli* expression system for the rapid purification and characterization of a v-abl tyrosine protein kinase. *Oncogene Research* 5 (1990): 161–173.

Nishizuka, Y. The role of protein kinase C in cell surface signal transduction and tumour promotion. *Nature* 308 (1984): 693–698.

Nowell, P. C., and D. A. Hungerford. A minute chromosome in human chronic granulocytic leukemia. *Science* 132 (1960): 1497.

———. Chromosome studies on normal and leukemic human leukocytes. *Journal of the National Cancer Institute*. 25 (1960) : 85–109.

O'Brien, S. G., F. Guilhot, and R. A. Larson et al.; IRIS investigators. Imatinib compared with interferon and low-dose cytarabine for newly diagnosed chronic-phase chronic myeoid leukemia. *New England Journal of Medicine* 348 (2003): 994–1004.

Rabstein, L. S., A. F. Gazdar, H. C. Chopra, and H. T. Abelson. Early morphological changes associated with infection by a murine nonthymic lymphatic tumor virus. *Journal of the National Cancer Institute* 46 (1971): 481–491.

Rowley, J. D. Chromosomal patterns in myelocytic leukemia. *New England Journal of Medicine* 289 (1973): 220–221.

———. Letter: A new consistent chromosomal abnormality in chronic myelogenous leukemia identified by quinacrine fluorescence and Giemsa staining. *Nature* 243 (1973): 290–293.

Sawyers, C. L., A. Hochhaus, and E. Feldman. Imatinib induces hematologic and cytogenetic responses in patients with chronic myeloid leukemia in myeloid blast crisis: Results of a phase II study. *Blood* 99 (2002): 3530–3539.

Schindler, T., W. Bornmann W, P. Pellicena, W. T. Miller, B. Clarkson, and J. Kuriyan. Structural mechanism for STI-571 inhibition of Abelson tyrosine kinase. *Science* 289 (2000): 1938–1942.

Stam, K., N. Heisterkamp, G. Grosveld, A. de Klein, R. S. Verma, M. Coleman, H. Dosik, and J. Groffen. Evidence of a new chimeric *bcr/c-abl* mRNA in patients with chronic myelocytic leukemia and the Philadelphia chromosome. *New England Journal of Medicine* 313 (1985): 1429–1433.

Talpaz, M, R. T. Silver, and B. J. Druker et al. Imatinib induces durable hematologic responses and cytogenetic responses in patients with accelerated phase chronic myeloid leukemia: Results of a phase 2 study. *Blood* 99 (2002): 1928–1937.

Tamaoki T, H. Nomoto, I. Takahashi, Y. Kato, M. Morimoto, and F. Tomita. Staurosporine, a potent inhibitor of phospholipid/Ca++dependent protein kinase. *Biochemical and Biophysical Research Communications* 135 (1987): 397–402.

Whang, J., E. Frei III, J. H. Tjio, P. P. Carbone, and G. Brecher. The distribution of the Philadelphia chromosome in patients with chronic myelogenous leukemia. *Blood* 22 (1963): 664–673.

Witte, O. N. Involvement of the *abl* oncogene in human chronic myelogenous leukemia. Oncogenes and Cancer. Utrecht: *Japan Society Press, 1987. S. A. Aaronson et al, editors. 143-149.*

———. Role of the BCR-ABL oncogene in human leukemia. *Cancer Research* 53 (1993): 485–489.

Witte, O. N., A. Dasgupta, and D. Baltimore. Abelson murine leukemia virus protein is phosphorylated in vitro to form phosphotyrosine. *Nature* 283 (1980): 826–831.

Witte, O. N., N. E. Rosenberg, and D. Baltimore. Identification of a normal cellular protein cross-reactive to the major Abelson murine leukaemia virus gene product. *Nature* 281 (1979): 396–398.

Witte, O. N., N. Rosenberg, M. Paskind, A. Shields, and D. Baltimore. Identification of an Abelson murine leukemia virus-encoded protein present in transformed fibroblast and lymphoid cells. *Proceedings of the National Academy of Sciences of the United States of America* 75 (1978): 2488–2492.

Young, J. C., and O. N. Witte. Selective transformation of primitive lymphoid cells by the BCR/ABL oncogene expressed in long-term lymphoid or myeloid cultures. *Molecular and Cellular Biology* 8 (1988): 4079–4087.

Zimmermann, J., E. Buchdunger, H. Mett, T. Meyer, and N. B. Lydon. Potent and selective inhibitors of the Abl-kinase: Phenylamino-pyrimidine (PAP) derivatives. *Bioorganic and Medicinal Chemistry Letters* 7 (1997): 187–192.

OTHER PUBLISHED WORK

These articles, books, and other publications provide summaries of research, historical perspectives, and other references. Many of these works are review articles, papers that synthesize the most pertinent original research in a given field. Additional reference materials were used as sources for this book, but the following represent the most key reports.

Arnold, K. After 30 years of laboratory work, a quick approval for STI571. *Journal of the National Cancer Institute* 93 (2001): 972–973.

Baltzer, F. "Theodor Boveri: The Life of a Great Scientist 1862–1915." Berkeley: University of California Press, 1967. Available online at http://9e.devbio.com/article.php?ch=2&id=25.

Bazell R. *HER-2*. New York: Random House, 1998.

Bishop, J. M. Oncogenes. *Scientific American* 246 (1982): 68–78.

Cohen, P. Protein kinases—the major drug targets of the twenty-first century? *Nature Reviews Drug Discovery* 1 (2002): 309–315.

Druker, B. J. Translation of the Philadelphia chromosome into therapy for CML. ASH 50th anniversary review. *Blood* 112 (2008): 4808–4817.

Druker, B. J., C. L. Sawyers, R. Capdeville, J. M. Ford, M. Baccarani, and J. M. Goldman. Chronic myelogenous leukemia. *Hematology*. American Society of Hematology Education Program (2001): 87–112.

Hunter, T. The proteins of oncogenes. *Scientific American* 251 (1984): 70–79.

———. Treatment for chronic myelogenous leukemia: The long road to imatinib. *Journal of Clinical Investigation* 117 (2007): 2036–2043.

Hunter, T., and W. Eckhart. The discovery of tyrosine phosphorylation: It's all in the buffer! *Cell* S116 (2004): S35–S39.

Kharas, M. G., and G. Q. Daley. From hen house to bedside: Tracing Hanafusa's legacy from avian leukemia viruses to SRC to ABL and beyond. *Genes & Cancer* 1 (2011): 1164–1169.

Kurzrock, R., H. M. Kantarjian, B. J. Druker, and M. Talpaz. Philadelphia chromosome–positive leukemias: From basic mechanisms to molecular therapeutics. *Annals of Internal Medicine* 138 (2003): 819–830.

Langreth, R. Big bucks. *Forbes,* May 21, 2007.

Lawce, H. Genetic technology and CML: Culture and history. *Journal of the Association of Genetic Technologists* 37 (2011): 29–30.

Lydon, N. B. Attacking cancer at its foundation. *Nature Medicine* 15 (2009): xix–xxiii.

Lydon, N.B., and B. J. Druker. Lessons learned from the development of imatinib. *Leukemia Research* 28, supplement 1 (2004): S29–S38.

Monmaney T. A triumph in the war against cancer. *Smithsonian,* May 2011.

Mueller, J. M. Taking TRIPS to India—Novartis, Patent Law, and Access to Medicines. *New England Journal of Medicine* 256 (2007): 541–543.

Mukherjee, S. *The Emperor of All Maladies.* New York: Scribner, 2010.

Nowell, P. C. Genetic alterations in leukemias and lymphomas: Impressive progress and continuing complexity. *Cancer Genetics and Cytogenetics* 94 (1997): 13–19.

Pollack, A. Therapies for cancer bring hope and failure. *New York Times,* June 15, 2010.

Rosenberg, N., and K. Beemon. Mechanisms of Oncogenes by Avian and Murine Retroviruses. In: *Current Cancer Research.* New York: Springer Science + Business Media, 2012.

Rous, P. Nobel Lecture, 1966. Available online at http://www.nobelprize.org/nobel_prizes/medicine/laureates/1966/rous-lecture.html.

Sharat Chandra, H., N. C. Heisterkamp, A. Hungerford, J. J. D. Morrissette, P. C. Nowell, J. D. Rowley, and J. R. Testa. Philadelphia Chromosome Symposium: Commemoration of the 50th anniversary of the discovery of the Ph chromosome. *Cancer Genetics* 204 (2011):171–179.

Temin, H. M. Mechanism of cell transformation by RNA tumor viruses. *Annual Review of Microbiology* 25 (1971): 609–648.

Vasella, D., and R. Slater. *Magic Cancer Bullet.* New York: HarperCollins, 2003.

US Department of Health and Human Services. Office of Inspector General. The Orphan Drug Act—Implementation and impact. May 2001. OEI-09-00-00380. Available online at http://www.dhhs.gov/progorg/oei.

US Food and Drug Administration. Summary basis of approval (SBA) for Gleevec. Document provided by FOI Services. Document number 5202527.

Wade, N. Powerful anti-cancer drug emerges from basic biology. *New York Times,* May 8, 2001.

Weiss, R. A., and P. K. Vogt. 100 years of RSV. *Journal of Experimental Medicine* 208 (2011): 2351–2355.

Wong S., and O. N. Witte. The BCR-ABL story: Bench to bedside and back. *Annual Review of Immunology* 22 (2004): 247–306.

ACKNOWLEDGMENTS

Thanks to all of the scientists and clinicians who agreed to interviews, including Brian Druker, Naomi Rosenberg, Owen Witte, David Baltimore, Nick Lydon, Alex Matter, Daniel Vasella, Charles Sawyers, Steve Goff, Sir Philip Cohen, Ray Erickson, Herb Abelson, Michael Mauro, Peter Nowell, Janet Rowley, Jürg Zimmermann, Elisabeth Buchdunger, Helen Lawce, Renaud Capdeville, John Goldman, Peter Traxler, Moshe Talpaz, Brian Hemmings, Peter Parker, Emil Freireich, George Daley, J. Michael Bishop, Nora Heisterkamp, Kara Johnson, and Joel Crouch. Additional thanks go to Steve Goff and Helen Lawce for reading and commenting on early drafts. I thank Gary Eichner, Judy Orem, Frank Orem, Alice Hungerford, Hans Loland, Beverly Alex Owens, Kelly Mitchell, Suzan McNamara, Alexandra Hardy, and Jennifer Gangloff for sharing their stories and allowing me a glimpse into their personal lives. Thanks also to LaDonna Lopossa, Dori Mortenson, Jay Weinstein, and Virginia Garner. I also want to recognize all those who contributed to the *STI Gazette* and the "Appreciation Album," both of which were source material for this book.

To all of these individuals, I hope I have shown respect in walking the hallowed ground of other people's lives.

This book emerged through the encouragement and support of many people. At my publisher, The Experiment, those people are Matthew Lore and Nicholas Cizek. Thanks also to Russell Galen, literary agent. I'm grateful to Sarah Bowden and Elisa Williams at OHSU and Sarah Kestenbaum at Ruder Finn for their assistance, and to Norma McLemore for copyediting and Jason Rothauser for proofreading. Steve Kurlander, Paul McDaniel, Tanya McKinnon, Siena Siegel, and Joy Pincus were also all part of this book coming to life.

I thank my family for giving me space, inspiration, and comic relief.

I have done my best to tell this story accurately, referring to the published literature to check the science and the data and, as much as possible, cross-checking each person's memory of the events described in this book. Any errors, omissions, or oversimplifications are my own and should not be misconstrued as a reflection on anyone mentioned above.

INDEX

C

PHOTO CREDITS

1. Peter Nowell, MD, and David Hungerford: Photograph taken by Larry Keighley, courtesy of Alice Hungerford.

2, 3. Microscope photographs of the Philadelphia chromosome: Photograph courtesy of Alice Hungerford.

4. Naomi Rosenberg's cell cultures: Photograph courtesy of Naomi Rosenberg.

5. Barred Plymouth Rock hen: 1910, Rockefeller University Press. Originally published in *The Journal of Experimental Medicine.* 12:696–705.

6. Bone marrow biopsy: Republished with permission of the American Society for Clinical Investigation, from "Applying the Discovery of the Philadelphia chromosome," Daniel W. Sherbenou and Brian J. Druker, *Journal of Clinical Investigation*, Volume 117, Issue 8, 2007; permission conveyed through Copyright Clearance Center, Inc.

7. *src* probe diagram: Illustration © Molly Feuer, Feuer Illustration. Text by the author.

8. Janet D. Rowley, MD: Photograph courtesy of Janet D. Rowley.

9. Philadelphia chromosome translocation: © 2007 Terese Winslow, U.S. Govt. has certain rights.

10. Karyotype with Philadelphia chromosome: From the Department of Pathology and Clinical Laboratory of the University of Pennsylvania School of Medicine. Image courtesy of Peter C. Nowell, MD, and Kristin Nowell.

11. From Philadelphia chromosome to CML diagram: Illustration © Molly Feuer, Feuer Illustration. Text by the author.

12, 13. Hematologic and cytogenetic (FISH) responses to CML: Republished with permission of the American Society for Clinical Investigation, from "Applying the Discovery of the Philadelphia chromosome," Daniel W. Sherbenou and Brian J. Druker, *Journal of Clinical Investigation*, Volume 117, Issue 8, 2007; permission conveyed through Copyright Clearance Center, Inc.

14. Jürg Zimmermann: Photograph courtesy of Jürg Zimmermann.

15. Elisabeth Buchdunger: Photograph courtesy of Elisabeth Buchdunger.

16. Warren Alpert Prize award ceremony: Photograph courtesy of The Warren Alpert Foundation.

17. Brian Druker, MD, and LaDonna Lopossa: Photograph by Michael McDermott and courtesy of Oregon Health & Science University.

18. Gleevec pill: Photograph courtesy of Novartis.

19. 50th anniversary group photo: Photograph courtesy of Fox Chase Cancer Center.

20. Gary Eichner and son: Photograph courtesy of Gary Eichner.

21. Novartis ad featuring Suzan McNamara: © Novartis.

ABOUT THE AUTHOR

Jessica Wapner is a freelance science writer focused mainly on health care and medicine. Her work has appeared in publications including *Scientific American*, *Slate*, *The New York Times*, theatlantic.com, *New York*, *Science*, *Nature Medicine*, the *Ecologist*, the *Scientist*, and *Psychology Today*. Her writing on cancer research and treatment has also appeared in the patient-focused magazines *CR* and *Cure*, and she has been a frequent contributor to the industry publication *Oncology Business Review*. She lives with her family in Beacon, New York.